Computational Complexity of Sequential and Parallel Algorithms

WILEY SERIES IN COMPUTING

Consulting Editor
Professor D. W. Barron
Department of Computer Studies, University of Southampton,
UK

Computational Complexity of Sequential and Parallel Algorithms

Lydia Kronsjö
The Centre for Computing and Computer Science
The University of Birmingham

A Wiley–Interscience Publication

JOHN WILEY & SONS
Chichester · New York · Brisbane · Toronto · Singapore

Library of Congress Cataloging in Publication Data:

Kronsjö, Lydia.
    Computational complexity of sequential and
parallel algorithms.

    (Wiley series in computing)
    Includes index
    1. Electronic digital computers—Programming.
2. Mathematics—Data processing. 3. Algorithms.
4. Parallel processing (Electronic computers)
5. Sequential processing (Computer science)
I. Title. II. Series.
QA76.6.K763   1986   001.64'2   85–9566

ISBN 0 471 90814 2

British Library of Cataloguing in Publication Data:

Kronsjö, Lydia
    Computational complexity of sequential and
    parallel algorithms.—(Wiley series in computing)
    1. Algorithms
    I. Title
    511'.8   QA9.58

ISBN 0 471 90814 2

Printed and bound in Great Britain

Contents

Preface

In the 1930s the notion of an algorithm was formulated as a precise mathematical concept. The Turing machine was conceived and with remarkable simplicity captured the mechanism of problem solving; for the first time the notion of an algorithmically computable function was formalized. Now, 50 years later, the science of computing is a well-established and vigorously researched field; we have algorithms which solve large and complex problems on our powerful computers in seconds, we are aware of the existence of those problems which cannot be solved by a computer, and we have encountered problems which we cannot solve in realistic time limits in spite of having very powerful computers. Until recently, algorithms were constructed under the assumption that a computer executes the algorithm's instructions in sequential order, one after another. However, the latest advances in computer technology have brought about a new generation of computing machines—parallel computers—which are able to perform several computing tasks simultaneously, and all the indications are that the future belongs to these new machines because they offer computing power which is higher than any sequential machine may physically be capable of. For the science of algorithms it means that a new, conceptually different, approach is needed in the design of algorithms. At the same time, undoubtedly, the wealth of knowledge which has been acquired by the era of sequential algorithms should prove very useful in the new developments.

This book consists of two parts. In Part One some conceptually important achievements in the design methodology of sequential algorithms are presented. In Part Two new solutions in the field of parallel algorithms are introduced. Whenever possible both sequential and parallel algorithms for the same class of problems are treated. The author's purpose was to present a concise treatment of the important results from the theory and applications of sequential algorithms and, in parallel, give an introduction to the fast developing area of parallel algorithms. Such presentation of material in one book is aimed at giving the reader an opportunity to compare, contrast, and appreciate unique features of the two kinds of computing environment. A comprehensive list of references is included which may be of service to the reader intending to go further.

The book can be recommended for second- and third-year undergraduate and postgraduate students in mathematics, computer science, and software engineering. Chapters 1 and 9 are introductory to Parts One and Two respectively; apart from these chapters, the book can be followed in any selective order as the chapters are essentially self-contained.

I would like to warmly acknowledge the work of several of my students who contributed to the material of the book by developing examples which enhanced the algorithmic exposition and, in some cases, by devoting detailed analysis to some arguments of our discussion. Among them, Martin Edge worked with Pan's algorithms, Matthew Levin carried out an excellent project on asynchronous algorithms for solving non-linear systems, Nilden Eminer researched into echo algorithms, and Stephen Lee produced a crisp project on parallel sorts. My colleague Dennis Parkyn kindly read most of the manuscript. The sustained interest and encouragement of my close friends is most gratefully appreciated. To my son Tim, husband Tom, and his father Erik go the last but probably deepest thank you for their always being close and understanding.

LYDIA KRONSJÖ
Birmingham, 2 February 1985

Part One
Sequential Algorithms

1

Introduction

Computational complexity addresses itself to the quantitative aspects of the solution of computational problems. Its concern is with the development of techniques for bounding the amount of computational resources which are necessary and sufficient to solve certain classes of problems algorithmically.

A typical problem in computational complexity may be characterized as follows.

We have a class of similar computational tasks, a problem P, e.g. inversion of matrices, parsing of strings, sorting of files, where the parameter n is a measure of an instance of P, which can be the order of a matrix to be inverted, or the length of a string to be parsed, or the size of the file to be sorted in order; a measure typically 'measures' the number of inputs, x_1, x_2, $\ldots$, x_n.

We also have a basic set of operations for carrying out the computation task in hand, e.g. the set of arithmetic operations $+$, $-$, $*$, $/$, the operation of comparison of two elements, the operation of fetching a record from computer memory, etc.

A step-by-step procedure for solving the problem is called an algorithm. A typical pattern of research in computational complexity is the design and development of algorithms which solve the problem with a reduced number of basic operations and the establishment of arguments which allow us to determine bounds on the possible extent of that reduction. The latter is a particularly difficult problem, and for many present-day problems a gap remains between the requirements of the existing algorithms and those of the minimum bounds for solving the problem.

At a certain level of detail we can usefully distinguish between the algorithm—an abstraction describing a process of computation that can be implemented on many computers—and the computer program, which is an implementation of the algorithm for a particular machine. Algorithms are abstractions whose generality is intended to transcend the specifics of any implementation. We shall not develop Pascal programs, but rather strive to describe algorithms in a form unencumbered by any preordained syntax or semantics, since the algorithms are aimed at thinking readers rather than computers.

There are often several algorithms which solve P, and we may ask the following questions:

3

(a) What is a 'good' algorithm for solving P?

To answer this question we may seek to compare the relative efficiencies of different algorithms.

(b) What is the minimum number of basic operations required to solve P?

We may then seek to find specific algorithms that solve the problem with the minimum number of operations.

(c) Is the problem perhaps such that no algorithm will solve it in a practically reasonable time?

The problems with a positive answer to this question are called *intractable*. The idea of an intractable problem was independently introduced by Cobham (1964) and Edmonds (1965). It helped to divide all problems for which there exist algorithms, into two distinct classes: efficient or practically solvable algorithms which usually correspond to some significant structure in the problem, and inefficient algorithms which often reflect a brute-force search and whose execution time explodes into exponential and superexponential growth.

In order to make an informed choice between algorithms which solve the same problem, systematic information about the algorithms' performance is needed.

1.1 Measures of Complexity

One way to compare the algorithms' performance is to compute some measure of their efficiency. To be useful, this measure should be machine independent. Good algorithms tend to remain good if they are expressed in different programming languages or run on different machines.

The two most useful measures are the time required to execute the algorithm and the memory needed by the algorithm. These measures are generally expressed as a function of the problem size.

The time complexity function is the computational time requirements which in practice are often calculated as the number of times a particular computer operation occurs, e.g.

(i)   **for** $i := 1$ **to** n **do**

$\quad\quad x_i := x_i + 1$ $\quad\quad\quad\quad$ is $O(n)$—read as 'of order n'.

(ii) **for** $i := 1$ **to** n **do**

$\quad\quad$ **for** $j := 1$ **to** n **do**

$\quad\quad\quad x_{ij} := 2x_{ij}$ $\quad\quad\quad$ is $O(n^2)$.

The space complexity function is the space requirements of the algorithm. This may be for the storage of matrices, intermediate data, etc. and the function represents the peak amount required.

On the program level, other measures, such as the program length, which is indicative of computation time, and the depth of the program, i.e. the number of layers of concurrent steps into which the problem can be decomposed, are useful. Depth corresponds to the time that the program would require under parallel computation.

The time of the algorithm and storage space are important measures and are particularly appropriate if the algorithm (program) is to be run often. The time of the algorithm is the factor that restricts the size of problems which can be solved by computer and the program length in some sense measures the simplicity of an algorithm. Tarjan (1978) calls an algorithm a simple algorithm if it has a short program and a short correctness proof, and an elegant algorithm if it has a short program and a long correctness proof. The program length as a measure of the algorithm's efficiency is most appropriate if programming time is important or if the program is to be run frequently.

1.2 Computer Model

Any specific complexity measure, i.e. execution time, storage space, program length, program depth, assume some model of an effective 'step' in a computation process, that is, a computer model for measuring the computation time/space of an algorithm.

The fame of the first computer model is attributed to the celebrated Turing machine of the mid-1930s, which was conceived to be an automaton equipped with an infinite supply of paper tape marked off in square regions, and capable of just four actions: it could move the tape one square, left or right, it could place a mark in a square, it could erase a mark already present, and at the end of a calculation it could halt. These operations were to be performed according to a sequence of instructions built into the internal mechanism. The machine was a conceptual device for automatically solving problems in mathematics and logic. With his machine, Turing for the first time rigorously formalized the notion of an algorithmically computable function.

Reduced to its essentials the Turing machine is a language for stating algorithms as powerful in principle as the more sophisticated languages now employed for communicating with computers. Several computer models have been proposed since. Following the Turing concept, many computer models assume a sequential (linear tape) storage media. The simplicity of such models makes them useful in theoretical studies of computational complexity, but serial storage media implies artificial restrictions from the point of view of efficient implementation of algorithms. Real computers have random access memories, so a better computer model would be a random-access machine. The question seems to be how better to conceive such a model.

One model of a random access machine (RAM), due to Cook and Reckhow (1973), is an abstraction of a general-purpose digital computer. The memory of the machine consists of an array of n words, each of which is able to hold a single integer. The storage words are numbered consecutively from 1 to n. The number of a storage word is its address.

The machine also has a fixed finite set of registers, each able to hold an integer. For problems involving real numbers, storage words and registers are allowed to hold real numbers. In one step, the machine can transfer the contents of a register to a storage word whose address is in a register, or transfer to a register the contents of a storage word whose address is in a register, or perform an arithmetic operation on the contents of two registers, or compare the contents of two registers.

A program of fixed finite length specifies the sequence of operations to be carried out. The initial configuration of memory represents the input data, and the final configuration of memory represents the output.

Various modifications and expansions to the basic model were introduced by Aho *et al.* (1974). Other machine models were suggested by Kolmogorov (1953), Kolmogorov and Uspenskii (1963), Knuth (1968), Schönhage (1973), and Tarjan (1977). Knuth, Schönhage and Tarjan first developed and subsequently improved the so-called pointer machine. As distinct from a RAM machine (which has a finite fixed memory and a set of registers' structure) a pointer machine consists of an extendable collection of nodes, each divided into a fixed number of named fields. A field holds a number or a pointer to a node. In order to carry out an operation on the field in a node, the machine must have in a register a pointer to the node. The operations which the machine can perform are fetching from a node field, storing into a node field, creating a node, and destroying a node; but there is no address arithmetic on a pointer machine and hence algorithms that require such arithmetic, e.g. hashing, cannot be implemented on such a machine. A pointer machine is a less powerful model than a RAM machine, but nevertheless it is useful for the lower bound studies of a great many algorithms, notably for the list processing algorithms (Tarjan, 1983).

A more recent Schönhage's storage modification machine (Schönhage, 1980) is organized in such way that it can be viewed either as a Turing machine that builds its own storage structure or as a unit cost (step) RAM machine that can only copy, add, or subtract one, or store or retrieve in one step.

All machine models described share two properties: they carry out one step at a time, i.e. they are sequential, and the future behaviour of the machine is uniquely determined by its present configuration. In later sections we shall discuss extensions to these machine models which incorporate the concepts of non-determinism and parallel computation. Non-deterministic machine models are particularly useful in theoretical studies of computational complexity (Aho *et al.*, 1974; Garey and Johnson, 1979). On the other hand, the novel machine architectures which have been made possible by very large scale integration (VLSI) give rise to a new type of algorithms, parallel algorithms, which are becoming more and more important.

Part One of the book deals with the ideas and achievements in the field of sequential algorithms, stressing the importance of non-determinism in coping with computationally difficult problems. In Part Two we discuss several problems for which efficient and interesting parallel algorithms have been developed.

2

Arithmetic Complexity of Computations

The arithmetic complexity of computations concerns the algorithms where the set of basic operations required for carrying out the task of computing the solution consists of arithmetic operations, $+$, $-$, $*$, $/$, and perhaps extended to include $\sqrt{}$, max or min.

The two major questions of arithmetic complexity—what is the minimum number of arithmetic operations needed to perform the computation and how can we obtain a better algorithm when improvement is possible?—pertain to any computation which may be carried out using the set or subset of arithmetic operations, $+$, $-$, $*$, $/$.

The major classes of arithmetic problems include algebraic processes, such as solution of linear systems, matrix multiplication and inversion, evaluation of the determinant, evaluation of a polynomial, at a point or at n points, and evaluation of the polynomial derivatives, iterative computations, e.g. the root-finding problem, solution of linear systems, and multiplication of two integers.

More recent problems in arithmetic complexity are explicitly concerned with the binary representation of a problem within a computer and relate the requirements of the numerical accuracy of computations. Pan (1980) has formulated the following matrix problem. The entries of matrices are numbers given with a certain precision p in binary form. Find the lower and upper bounds on the number of bit operations involved in the evaluation (with another given precision q) of the product of the two matrices.

2.1 Discoveries of the 1960s

The 1960s are one of the major milestones in the history of computational complexity. During this decade three very surprising algorithms were discovered—for the multiplication of two integers, for computing the discrete Fourier transform, and for the product of two matrices. As Winograd (1980) has aptly noted, all three computational problems have been known for a very long time, and the discovery of more efficient algorithms for their execution was an exciting result which encouraged further investigations of the limits on the efficiency of algorithms.

Today, the importance of these discoveries is not only historic; the new algorithm for computing the Fourier transform has had indeed a profound

7

impact on the way many computations are being performed. Perhaps one of the most remarkable uses of the Fast Fourier transform (FFT) algorithm is in medical analysis, the process known as computerized tomography. We shall briefly outline the three historical results.

2.2   Product of Integers

In 1962 two Russian mathematicians, Karatsuba and Ofman, published a paper in which a new way was given for multiplying two large integers. Using the usual method, as we know it, the product of two n-digit numbers is obtained by performing n^2 multiplications of single digits and about the same number of single-digit additions. The method of Karatsuba and Ofman computes the product in $O(n^{\log 3})$ time and is based on the following idea.

Let x and y be two n-digit numbers, where $n = 2m$, an even number. If b denotes the base then

$$x = x_0 + x_1 b^m, \qquad y = y_0 + y_1 b^m , \qquad (2.2.1)$$

where x_0, x_1, y_0, and y_1 are m-digit numbers. The product w is then expressed as

$$\begin{aligned} z = xy &= (x_0 + x_1 b^m)(y_0 + y_1 b^m) \\ &= x_0 y_0 + (x_0 y_1 + x_1 y_0)b^m + x_1 y_1 b^{2m} \end{aligned} \qquad (2.2.2)$$

and its computation can be viewed as consisting of four steps: (1) compute $x_0 y_0$; (2) compute $x_0 y_1 + x_1 y_0$; (3) compute $x_1 y_1$; (4) perform $n = 2m$ single-digit additions.

Multiplication by the power of 2 in the binary environment represents a simple shifting operation and is ignored in this analysis. Thus $4M$ (M = multiplication) of m-digit numbers and $2A$ (A = addition) of $2m$-digit numbers is required to obtain w.

This is the same as the estimate on the number of operations in the usual method. The key to the new algorithm is the way in which $x_0 y_0$, $x_0 y_1 + x_1 y_0$, and $x_1 y_1$ are computed. Their computation is based on the identities

$$\begin{aligned} x_0 y_0 &= x_0 y_0, \\ x_0 y_1 + x_1 y_0 &= (x_0 - x_1)(y_1 - y_0) + x_0 y_0 + x_1 y_1, \\ x_1 y_1 &= x_1 y_1, \end{aligned} \qquad (2.2.3)$$

from which it follows that to obtain the left-hand-side values of the identities one needs to find three products of two m-digit numbers and carry out some additions and subtractions. If the addition or subtraction of two single-digit numbers with the possibility of carry or borrow is taken as a unit of addition then the product of two n-digit numbers uses $3M$ of m-digit numbers and $4n$ units of addition.

The same scheme can be used to obtain each of the products $x_0 y_0$, $(x_0 - x_1)(y_0 - y_1)$, and $x_1 y_1$. This leads to the result that for $n = 2^s$, a power of 2, the product of two $n = 2^s$ digit numbers is obtained using $3^s M$ of single-digits and $8(3^s - 2^s)$ units of addition, assuming the initial condition that for $m = 2^0 = 1$ only one single digit M and no units of addition are needed.

When n is not a power of 2, the numbers can be 'padded' by adding enough leading zeros. So the formulae are still valid if we take $s = \lceil \log_2 n \rceil$ for any n.

In summary, the method shown for computing the product of two n-digit numbers uses at most $3*3^{\log_2 n} = 3n^{\log_2 3} = 3n^{1.59}$ single-digit M and $8*3*3^{\log_2 n} = 24n^{\log_2 3} = 24n^{1.59}$ units of addition.

The more detailed analysis can reduce the constants 3 and 24, but the order of the function on the number of operations will remain $n^{\log_2 3}$. Since the Karatsuba and Ofman paper the result has been improved by several authors. The currently fastest asymptotic performance time for the number product is $O(n \log n \log \log n)$ and was devised on a multitape Turing machine by Shönhage and Strassen (1971) using FFT.

2.3 The Fast Fourier Transform

Given is a set of n points

$$a_0, a_1, a_2, \ldots, a_{n-1}.$$

The discrete Fourier transform (DFT) on these points is a set of n points

$$A_0, A_1, A_2 \ldots, A_{n-1},$$

where A_p's are computed by the formula

$$A_p = \sum_{q=0}^{n-1} w^{pq} a_q, \qquad p = 0, 1, \ldots, n-1, \tag{2.3.1}$$

and where w is the nth root of unity, that is $w = e^{2\pi i/n}$, $i = \sqrt{-1}$.

Straightforward computation of each A_p uses $n-1$ complex M and $n-1$ complex A, thus yielding a computational procedure of complexity $O(n^2)$.

In 1965 Cooley and Tukey observed that whenever $n = rs$ is a composite number a more efficient method of computation exists. Each p in $0 \leqslant p \leqslant n-1$ can be written uniquely as

$$p = p_1 + p_2 r, \qquad 0 \leqslant p_1 < r, \qquad 0 \leqslant p_2 < s, \tag{2.3.2}$$

an each q in $0 \leqslant q \leqslant n-1$ as

$$q = q_1 s + q_2, \qquad 0 \leqslant q_1 < r, \qquad 0 \leqslant q_2 < s. \tag{2.3.3}$$

Therefore

$$A_{p_1 + p_2 r} = \sum_{q_1=0}^{r-1} \sum_{q_2=0}^{s-1} w^{(p_1 + p_2 r)(q_1 s + q_2)} a_{q_1 s + q_2}$$

$$= \sum_{q_1=0}^{r-1} \sum_{q_2=0}^{s-1} w^{p_1 q_1 s}\, w^{p_1 q_2}\, w^{p_2 q_1 rs}\, w^{p_2 r q_2}\, a_{q_1 s + q_2}$$

$$= \sum_{q_1=0}^{r-1} \sum_{q_2=0}^{s-1} (w^s)^{p_1 q_1}\, w^{p_1 q_2}\, (w^r)^{p_2 q_2} a_{q_1 s + q_2},$$

since

$$w^{rs} = \overset{\circ}{w}{}^{n} = 1,$$

$$= \sum_{q_2=0}^{s-1} (w^r)^{p_2 q_2} \left[w^{p_1 q_2} \sum_{q_1=0}^{r-1} (w^s)^{p_1 q_1} a_{q_1 s + q_2} \right]. \tag{2.3.4}$$

Hence to compute each A_p we need to

(i) Compute the DFT on r points,

$$b_{p_1, q_2} = \sum_{q_1=0}^{r-1} (w^s)^{p_1 q_1} a_{q_1 s + q_2}. \tag{2.3.5}$$

(ii) Compute the factor

$$c_{p_1, q_2} = w^{p_1 q_2} b_{p_1, q_2}, \tag{2.3.6}$$

(iii) Compute the DFT on s points,

$$A_p = \sum_{q_2=0}^{s-1} (w^r)^{p_2 q_2} c_{p_1, q_2}. \tag{2.3.7}$$

This requires $(r+s-1) M$ and $(r+s-2) A$, and to compute n values A_p we shall need $n(r+s-1) M$ and $n(r+s-2) A$.

If r and s are themselves composite numbers we can use the same idea to compute the DFT of r or s points. In particular, when $n = 2^s$ is a power of 2 this computation process uses $2^s s = n \log_2 n \, M$ and $2^s (s-2) = n(\log_2 n - 2) A$. Thus the new algorithm, known as the FFT, is of complexity $O(n \log n)$, a staggering reduction as compared with $O(n^2)$.

As can be observed, the FFT takes advantage of the periodic nature of sine and cosine functions, i.e. $w^{\alpha i} = \sin\alpha + i \cos\alpha$, to greatly reduce the number of multiplications required in evaluating the DFT.

The tremendous reduction in computing time of the DFT which the FFT algorithm offers makes the DFT one of the most powerful mathematical tools used in the solution of many important practical problems. Of all the applications, perhaps the greatest effect on the world at large has been in the area of diagnostic medicine: the so-called computerized tomography (CT) has revolutionized radiology.

The Computerized Axial Tomography (CAT) Scanning and the FFT

In CT the FFT algorithm is used in reconstruction of an image of human body cross-section from data collected by measuring the intensity of X-ray beams transmitted through the cross-section, see Fig. 2.3.1(a). A tomogram is a picture of a slice—a display of an anatomical plane sectioning the body at a given orientation. In the CT method the information from different views

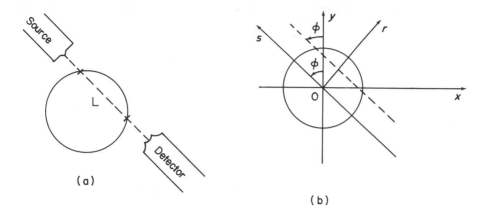

Figure 2.3.1 (a) The source-detector complex moves parallel to line L until the whole cross-section is 'X-rayed'. Next the position of the complex is changed by rotating it at some angle relative to line L and the process of X-raying the whole cross-section is repeated. (b) The X-ray attenuation at a point inside the cross-section is described in terms of an (x,y) co-ordinate system and the X-ray passing through the cross-section is described in an (r, s) coordinate system

within a single cross-sectional plane is acquired and subsequently processed to present a recognizable image. The CT scanner measures the attenuation of X-rays along a large number of lines through the cross-section of the body.

Along a single line the total X-ray attenuation is determined by the linear attenuation coefficients at individual points on the line. The linear attenuation coefficient at a point is determined by the tissue occupying that point. Since different tissue/tumour types have different linear attenuation coefficients, knowledge of the distribution of linear attenuation coefficients in the cross-section provides the desired information or a 'picture' of the cross-section.

In mathematical terms, the problem of the cross-section reconstruction can be formulated as follows. There is an unknown two-dimensional distribution of some physical parameter. A finite number of line integrals of this parameter is estimated from physical measurements. One wishes then to reconstruct the original distribution. Note that some inevitable simplifying conditions, as in any mathematical model abstracted from a physical process, are assumed in the computational procedure for reconstruction of a cross-section, e.g. a cross-section is assumed to be infinitely thin, i.e. it is a plane; both the X-ray source and the detector are assumed to be infinitely small, i.e. they are points, and the line between them is assumed to lie in the plane of the cross-section; the physical measurements provide the total X-ray attenuation along the line between the source and the detector.

Mathematical Derivation of the Method

In order to describe points in the cross-section we introduce an (x, y) co-ordinate system, as shown in Fig. 2.3.1(b) and denote the linear attenuation coefficient at point (x, y) by $f(x, y)$; this function is also known as the density function.

The value of this function outside the cross-section is set to zero on the assumption that the linear attenuation coefficient of air is zero. To describe the X-rays passing through the cross-section, we introduce another co-ordinate system, (r, s) which is rotated with respect to the (x, y) system by the same angle as the ray. Each ray is thus specified by co-ordinates (r, ϕ) where ϕ is the angle of the ray with respect to the y-axis and r is its distance from the origin. The co-ordinate s represents the path length along the ray.

The integral of the density function $f(x, y)$ along a ray (r, ϕ) is called the ray-sum or ray-projection, p:

$$p(r, \phi) = \int_{r, \phi} f(x, y) \, ds. \tag{2.3.8}$$

A complete set of ray-sums at a given angle is called a projection or profile. Thus the mathematical formulation of the cross-section reconstruction can be expressed as follows: given $p(r, \phi)$ for a large number of lines (r, ϕ), find $f(x, y)$ at selected points (x, y). The reconstruction methods based on direct solution of equation (2.3.8) are known as analytic reconstruction methods. They use the Fourier transform (FT) and the inverse Fourier transform (IFT). These methods, first derived in the 1950s, have become computationally viable only with the discovery of the time-saving FFT.

The Two-dimensional Fourier Reconstruction (Brooks and De Chiro, 1976)

The starting point in the derivation of the method is the representation of the density function as a two-dimensional Fourier integral:

$$f(x, y) = \int_{-\infty}^{\infty} \int_{-\infty}^{\infty} F(u, v) \, e^{2\pi i(ux+vy)} \, dudv. \tag{2.3.9}$$

By virtue of the definition the Fourier coefficients $F(u, v)$ are given by the IFT

$$F(u, v) = \int_{-\infty}^{\infty} \int_{-\infty}^{\infty} f(x, y) \, e^{-2\pi i(ux+vy)} dxdy \tag{2.3.10}$$

If we now rotate the (x, y)-axes to new axes (r, s), see Fig. 2.3.1(b), by the angle

$$\phi = \tan^{-1}(u/v) \tag{2.3.11}$$

and define

$$k = (u^2 + v^2)^{1/2}, \tag{2.3.12}$$

by a simple geometrical transformation, equation (2.3.10) can be rewritten as

$$F(u, v) = \int_{-\infty}^{\infty} \int_{-\infty}^{\infty} f(x, y)e^{-2\pi ikr}\, dr ds, \tag{2.3.13}$$

where $r = x \cos \phi + y \sin \phi$.

Next, exchanging the order of integration and noting that the s-integral is just the ray-projection $p(r, \phi)$, as given by equation (2.3.8), we get

$$F(u, v) = \int_{-\infty}^{\infty} p(r, \phi)\, e^{-2\pi ikr} dr = P(k, \phi), \tag{2.3.14}$$

where $P(k, \phi)$ is the FT of $p(r, \phi)$ with respect to r.

Equation (2.3.14) is of fundamental importance and is known as the projection theorem:

Each two-dimensional Fourier coefficient $F(u, v)$ of the density function $f(x, y)$ along a line through the origin in direction ϕ is equal to a corresponding one-dimensional Fourier coefficient $P(k, \phi)$ of the projection $p(r, \phi)$ taken at the right angle to the direction ϕ.

Replacing the Fourier integrals in equations (2.3.14) and (2.3.9) by the corresponding DFTs, the basic algorithm for reconstruction of cross-section can be formulated as follows:

(a) Collect a sufficient number of projections, $p(r, \phi)$;
(b) Compute the one-dimensional Fourier transform of projections, the coefficients $P(k, \phi)$;
(c) Compute the inverse two-dimensional transform, $f(x, y)$, noting that $F(u, v) = P(k, \phi)$, where $u/v = \tan \phi$ and $k = u/\cos \phi$.

The computational complexity of the procedure is dominated by the two-dimensional FT, which is of order $n^2(\log n)^2$, where n is the number of points in the DFT.

2.4   Matrix Product

The third new algorithm, for the product of two matrices, was discovered by Strassen in 1969.

$$\begin{bmatrix} A_{11} & A_{12} \\ A_{21} & A_{22} \end{bmatrix} \begin{bmatrix} B_{11} & B_{12} \\ B_{21} & B_{22} \end{bmatrix} = \begin{bmatrix} C_{11} & C_{12} \\ C_{21} & C_{22} \end{bmatrix}. \tag{2.4.1}$$

The obvious method for computing the product of two $n \times n$ matrices requires $n^3 M$ and $n^2(n-1) A$:

$$\begin{aligned}
\mathbf{C}_{11} &= \mathbf{A}_{11}\mathbf{B}_{11} + \mathbf{A}_{12}\mathbf{B}_{21}, & \mathbf{C}_{12} &= \mathbf{A}_{11}\mathbf{B}_{12} + \mathbf{A}_{12}\mathbf{B}_{22}, \\
\mathbf{C}_{21} &= \mathbf{A}_{21}\mathbf{B}_{11} + \mathbf{A}_{22}\mathbf{B}_{21}, & \mathbf{C}_{22} &= \mathbf{A}_{21}\mathbf{B}_{12} + \mathbf{A}_{22}\mathbf{B}_{22},
\end{aligned}$$

(2.4.2)

and thus is of complexity $O(n^3)$.

The new algorithm uses fewer operations if n is large enough. As was the case for the FFT algorithm and for the method for computing the product of two integers, Strassen's algorithm is also recursive in nature, that is, it reduces the problem of multiplying two $n \times n$ matrices to several instances of smaller problems.

Let $n = 2m$ be an even number. We partition an $n \times n$ matrix into four $m \times m$ matrices. In the matrix expressions (2.4.1) and (2.4.2) we can assume $\mathbf{A}_{ij}$'s, $\mathbf{B}_{ij}$'s and $\mathbf{C}_{ij}$'s are $m \times m$ matrices. The algorithm proceeds as follows.

Compute the seven matrices

$$\begin{aligned}
\mathbf{P}_1 &= (\mathbf{A}_{11} + \mathbf{A}_{22})(\mathbf{B}_{11} + \mathbf{B}_{22}), \\
\mathbf{P}_2 &= (\mathbf{A}_{21} + \mathbf{A}_{22})\mathbf{B}_{11}, \\
\mathbf{P}_3 &= \mathbf{A}_{11}(\mathbf{B}_{12} - \mathbf{B}_{22}), \\
\mathbf{P}_4 &= \mathbf{A}_{22}(\mathbf{B}_{21} - \mathbf{B}_{11}), \\
\mathbf{P}_5 &= (\mathbf{A}_{11} + \mathbf{A}_{12})\mathbf{B}_{22}, \\
\mathbf{P}_6 &= (\mathbf{A}_{21} - \mathbf{A}_{11}) (\mathbf{B}_{11} + \mathbf{B}_{12}), \\
\mathbf{P}_7 &= (\mathbf{A}_{12} - \mathbf{A}_{22}) (\mathbf{B}_{21} + \mathbf{B}_{22}),
\end{aligned}$$

(2.4.3)

and then compute

$$\begin{aligned}
\mathbf{C}_{11} &= \mathbf{P}_1 + \mathbf{P}_4 - \mathbf{P}_5 + \mathbf{P}_7, & \mathbf{C}_{12} &= \mathbf{P}_3 + \mathbf{P}_5, \\
\mathbf{C}_{21} &= \mathbf{P}_2 + \mathbf{P}_4, & \mathbf{C}_{22} &= \mathbf{P}_1 + \mathbf{P}_3 - \mathbf{P}_2 + \mathbf{P}_6.
\end{aligned}$$

(2.4.4)

Altogether, $7 M$ of $m \times m$ matrices and $18 A$ of $m \times m$ matrices are required.

The key to the economy of this algorithm is that it calls for only seven multiplications of $m \times m$ matrices rather than eight. Since the product of two $m \times m$ matrices uses $m^3 M$ and $m^2(m - 1) A$ and the sum of two $m \times m$ matrices uses $m^2 A$, the Strassen algorithm expends

$$7m^3 = (7/8)n^3 M$$

and

$$7(m^3 - m^2)+18m^2 \ 7m^3 + 11m^2 = (7/8)n^3+(11/4)n^2 A.$$

Whenever $n > 30$ we have both

$$(7/8)n^3 < n^3 \quad \text{and} \quad (7/8)n^3 + (11/4)n^2 < n^3-n^2.$$

Whenever m itself is an even number we can increase the number of arithmetic operations saved by using the same algorithm for computing each of the seven products of $m \times m$ matrices.

Taking $n = 2^s$ to be a power of 2 and denoting by $M(s)$ the number of multiplications needed to multiply two $n \times n$ matrices and by $A(s)$ the number of additions, we obtain

$$M(s+1) = 7M(s),$$
$$A(s+1) = 7A(s) + 18(4^s). \tag{2.4.5}$$

Using the initial conditions $M(0) = 1$ and $A(0) = 0$ we obtain

$$M(s) = 7^s = (2^s)^{\log_2 7} = n^{\log_2 7} = n^{2.807},$$
$$A(s) = 6(7^s - 4^s) = 6((2^s)^{\log_2 7} - (2^s))^2 = 6(n^{\log_2 7} - n^2). \tag{2.4.6}$$

In the case where n is not the power of 2, appropriate 'padding' of the two matrices by zeros is possible so as to make their dimension a power of 2. We then obtain $M(n) < 7n^{\log_2 7} - 42n^2$, where $M(n)$ denotes the number of multiplications needed to multiply two $n \times n$ matrices, and a similar result for $A(n)$. The Strassen algorithm thus requires only $4.7n^{2.807}$ operations.

Space Considerations for Strassen's Algorithm

Of the three algorithms considered, the Strassen method for computing the matrix product presents an example where extra space requirements are substantial due to the necessity to store the intermediate matrix components. At each level, k, of recursion in the algorithm a declaration of seven new auxiliary matrices of order $n/2^k$ is required and the eighth matrix is needed for the result. Hence the straightforward application of Strassen's method needs at least

$$n^2 + 8(n^2/4 + n^2/16 + \dots) = 11/3n^2$$

additional memory locations.

A version of the Strassen algorithm which performs *in situ* has been proposed by Kreczmar (1976). It reduces the extra storage requirements to $2/3n^2$ additional memory locations.

The Pan Algorithm

Since the publication of the Strassen method considerable work has been devoted to reducing the exponent of 2.807. In particular it was proved (Winograd, 1971; Hopcroft and Kerr, 1971) that two 2×2 matrices cannot be multiplied in less than seven multiplications. The next possibility then is to seek a further asymptotic speed-up by constructing fast algorithms which would multiply two 3×3 matrices in no more than $21M$, since for an algorithm to be faster than Strassen's it must have M such that $\log_3 M < \log_2 7$, where M is the number of multiplications used in multiplying two 3×3 matrices.

However, it proved difficult, if solvable at all, to multiply two 3×3 matrices in 21 or fewer multiplications. The closest result to 21 multiplications obtained so far is $23M$ and the result is due to Laderman (1976). Another algorithm, for multiplying two 5×5 matrices in $103M$ was proposed by Schachtel (1978). These results have suggested that perhaps still larger size matrices would yield a better outcome. A study in this direction by Pan (1978) has produced a new algorithm which appeared in 1978 and which is asymptotically faster than Strassen's.

To derive the new algorithm Pan used the fact that the evaluation of the product of $n \times n$ by $p \times m$ matrices and decomposing the trace of the product of three matrices, of size $n \times p$ by $p \times m$ by $m \times n$, are two equivalent problems. This fact is a particular case of the more general theorem stating that the evaluation of a set of bilinear forms in s multiplications and decomposing a trilinear form as a sum of s terms are two equivalent problems (Pan, 1972; Strassen, 1972). Example 2.4.1 illustrates this point.

Example 2.4.1

Consider the product of 3×2 and 2×3 matrices:

$$\begin{bmatrix} a_{11} & a_{12} & a_{13} \\ a_{21} & a_{22} & a_{23} \end{bmatrix} \times \begin{bmatrix} b_{11} & b_{12} \\ b_{21} & b_{22} \\ b_{31} & b_{32} \end{bmatrix} = \begin{bmatrix} c_{11} & c_{12} \\ c_{21} & c_{22} \end{bmatrix}, \tag{2.4.7}$$

and consider the bilinear form of the traditional algorithm,

$$c_{ij} = \sum_{k=1}^{3} a_{ik}b_{kj}, \qquad i, j = 1, 2, \tag{2.4.8}$$

where

$$c_{11} = a_{11}b_{11}+a_{12}b_{21}+a_{13}b_{31}, \qquad c_{12} = a_{11}b_{12}+a_{12}b_{22}+a_{13}b_{32},$$
$$c_{21} = a_{21}b_{11}+a_{22}b_{21}+a_{23}b_{31}, \qquad c_{22} = a_{21}b_{12}+a_{22}b_{22}+a_{23}b_{32}.$$

The equivalent representation in trilinear form is

$$\sum_{i,j,k} a_{ij}b_{jk}c_{ki}$$

$$\begin{aligned} &= a_{11}b_{11}c_{11} + a_{11}b_{12}c_{21} + a_{12}b_{21}c_{11} + a_{12}b_{22}c_{21} + a_{13}b_{31}c_{11} \\ &\quad + a_{13}b_{32}c_{21} + a_{21}b_{11}c_{12} + a_{21}b_{12}c_{22} + a_{22}b_{21}c_{12} + a_{22}b_{22}c_{22} \\ &\quad + a_{23}b_{31}c_{12} + a_{23}b_{32}c_{22} \\ &= (a_{11}b_{11} + a_{12}b_{21} + a_{13}b_{31})c_{11} + (a_{21}b_{11} + a_{22}b_{21} \\ &\quad + a_{23}b_{31})c_{12} + (a_{11}b_{12} + a_{12}b_{22} + a_{13}b_{32})c_{21} + (a_{21}b_{12} \\ &\quad + a_{22}b_{22} + a_{23}b_{32})c_{22}. \end{aligned} \tag{2.4.9}$$

In equation (2.4.9) the coefficient of c_{ij} is the (j, i)th element of the product C in equation (2.4.8).

There is more than one way in which the trilinear form can be represented. The Pan algorithm represents the trilinear form for computational purposes in such a way as to achieve reduction in the total number of multiplications required.

Example 2.4.2

Given are two 4×4 matrices

$$A = \begin{bmatrix} 5 & 3 & -1 & 4 \\ 2 & -5 & 7 & 9 \\ -7 & 4 & 0 & 5 \\ 8 & -3 & -4 & 1 \end{bmatrix} \quad \text{and} \quad B = \begin{bmatrix} 2 & 1 & -4 & -6 \\ 4 & 3 & -9 & 1 \\ 6 & 7 & 7 & 5 \\ -8 & 0 & 2 & 1 \end{bmatrix}.$$

Find their product, **C**.

Solution

Consider the trilinear form as

$$\sum_{i,j,k} a_{ij} b_{jk} c_{ki}$$

$$= \sum_{\substack{i+j+k \\ \text{is even}}} (a_{ij} + a_{k+1,i+1})(b_{jk} + b_{i+1,j+1})(c_{ki} + c_{j+1,k+1})$$

$$- \sum_{i,k} a_{k+1,i+1} \sum_{\substack{j:i+j+k \\ \text{is even}}} (b_{jk} + b_{i+1,j+1})c_{ki}$$

$$- \sum_{i,j} a_{ij} b_{i+1,j+1} \sum_{\substack{k:i+j+k \\ \text{is even}}} (c_{ki} + c_{j+1,k+1})$$

$$- \sum_{j,k} \sum_{\substack{i:i+j+k \\ \text{is even}}} (a_{ij} + a_{k+1,i+1})b_{jk}c_{j+1,k+1}.$$

Substituting for a_{ij}'s and b_{ij}'s and collecting the coefficients of c_{ij} we get:

$$\sum_{i,j,k} a_{ij} b_{jk} c_{ki} = -16c_{11} - 46c_{12} - 38c_{13} - 28c_{14} + 7c_{21}$$
$$+ 36c_{22} + 5c_{23} - 29c_{24} - 46c_{31} + 104c_{32} + 2c_{33}$$
$$- 31c_{34} - 28c_{41} + 27c_{42} + 51c_{43} - 70c_{44}.$$

Since the coefficient of c_{ij} is the (j, i)th entry in the matrix **C** we have

$$C = \begin{bmatrix} -16 & 7 & -46 & -28 \\ -46 & 36 & 104 & 27 \\ -38 & 5 & 2 & 51 \\ -28 & -29 & -31 & -70 \end{bmatrix}.$$

The number of multiplications required to calculate the product matrix **C** is 80, this is inferior to both Strassen's method (with 49 M) and the traditional method (with 64 M). However, when $n > 6$ the algorithm becomes faster than the traditional method.

Having studied various representations of the trilinear form and their efficiencies, Pan has suggested an algorithm which yields results superior to Strassen's method and is as follows:

$$\sum_{i,j,k} a_{ij}b_{jk}c_{ki} = T_0 - T_1 - T_2 - T_3, \qquad\qquad (2.4.10)$$

where

$$
\begin{aligned}
T_0 = \sum_{i,j,k \in S'(s)} \; &(a_{ij} + a_{jk} + a_{ki})(b_{jk} + b_{ki} + b_{ij})(c_{ki}+c_{ij}+c_{jk}) \\
&- (a_{ij} - a_{jk}^- + a_{ki}^-)(b_{jk}^- + b_{ki} - b_{ij}^-)(-c_{ki}^- + c_{ij} + c_{jk}) \\
&- (-a_{ij}^- + a_{jk}^- + a_{ki})(b_{jk} - b_{ki} + b_{ij}^-)(c_{ki} + c_{ij} - c_{jk}^-) \\
&- (a_{ij}^- + a_{jk} - a_{ki}^-)(-b_{jk}^- + b_{ki} + b_{ij})(c_{ki} - c_{ij}^{--} + c_{jk}^-) \\
&- (a_{ij}^- + a_{jk}^{--} - a_{ki}^-)(-b_{jk}^- + b_{ki} + b_{ij}^-)(c_{ki}^{--} - c_{ij} + c_{jk}^-) \\
&- (-a_{ij}^- + a_{jk} + a_{ki}^-)(b_{jk}^{--} - b_{ki} + b_{ij}^-)(c_{ki}^- + c_{ij}^{--} - c_{jk}^-) \\
&- (a_{ij}^{--} - a_{jk} + a_{ki}^-)(b_{jk}^{--} + b_{ki} - b_{ij}^-)(-c_{ki}^- + c_{ij}^{--} + c_{jk}^-) \\
&+ (a_{ij}^{--} + a_{jk}^- + a_{ki}^-)(b_{jk}^{--} + b_{ki} + b_{ij}^-)(c_{ki}^{--} + c_{ij} + c_{jk}^-),
\end{aligned}
$$

$$
\begin{aligned}
T_1 = \sum_{1 \le i,j \le s} \; &a_{ij}b_{ij}[(s-2w_{ij})c_{ij} + \Sigma^*(c_{ki} + c_{jk})] \\
&+ a_{ij}b_{ij}^- [(s-w_{ij})c_{ij} + \Sigma^*(-c_{ki}^- + c_{jk})] \\
&+ a_{ij}^- b_{ij}^- [(s-w_{ij})c_{ij} + w_{ji}c_{ji}^{--} + \Sigma^*(c_{ki}^- - c_{jk}^-)] \\
&+ a_{ij}^- b_{ij} [(s-w_{ij})c_{ij} - \Sigma^*(c_{ki} + c_{jk}^-)] \\
&+ a_{ij}^- b_{ij}^{--} [(s-w_{ij})c_{ij}^{--} - \Sigma^*(c_{ki}^{--} + c_{jk}^-)] \\
&+ a_{ij}^{--}b_{ij} [(s-w_{ij})c_{ij}^{--} - w_{ji}^{--}c_{ji} + \Sigma^*(c_{ki}^- - c_{jk}^-)] \\
&+ a_{ij}^{--}b_{ij}^- [(s-w_{ij})c_{ij} + \Sigma^*(-c_{ki}^- + c_{jk}^{--})] \\
&+ a_{ij}^{--}b_{ij}^- [(s-2w_{ij})c_{ij}^{--} + \Sigma^*(c_{ki}^{--} + c_{jk}^-)],
\end{aligned}
$$

$$
\begin{aligned}
T_2 = \sum_{1 \le i,j \le s} \; \{ &a_{ij} \Sigma^*(b_{ki} + b_{jk})c_{ij} - a_{ij} \Sigma^*(b_{ki} + b_{jk})c_{ij}^- \\
&+ a_{ij}^- \Sigma^*(b_{jk} - b_{ki}^-)c_{ij} + a_{ij}^- \Sigma^*[(b_{ki}^- - b_{jk}^-) - w_{ji}b_{ji}^-)c_{ij} \\
&+ a_{ij}^- [\Sigma^*(b_{ki}^- - b_{jk}^-) - w_{ji}^{--}b_{ji}]c_{ij}^- + a_{ij} \Sigma^*(b_{jk}^{--} - b_{ki}^-)c_{ij}^{--} \\
&- a_{ij}^{--} \Sigma^*(b_{ki}^{--} + b_{jk}^-)c_{ij}^- + a_{ij}^{--} \Sigma^*(b_{ki}^{--} + b_{jk}^-)c_{ij}^{--} \},
\end{aligned}
$$

$$
\begin{aligned}
T_3 = \sum_{1 \le i,j \le s} \; \{ &\Sigma^*(a_{ki} + a_{jk})b_{ij}c_{ij} + \Sigma^*[(a_{ki}^- - a_{jk}^-) - w_{ji}a_{ji}^-]b_{ij}^-c_{ij}^- \\
&- \Sigma^*(a_{ki} + a_{jk}^-)b_{ij}^-c_{ij} + \Sigma^*(a_{jk} - a_{ki}^-)b_{ij}c_{ij}^- \\
&+ \Sigma^*(a_{jk}^{--} - a_{ki})b_{ij}^{--}c_{ij}^- - \Sigma^*(a_{ki}^{--} + a_{jk})b_{ij}^-c_{ij}^{--} \\
&+ \Sigma^*[(a_{ki}^{--} - a_{jk}^-) - w_{ji}^-a_{ji}]b_{ij}c_{ij}^{--} + \Sigma^*(a_{ki}^{--} + a_{jk}^{--})b_{ij}^{--}c_{ij}^{--}.
\end{aligned}
$$

Here
$$
\begin{aligned}
S'(s) &= S_1'(s) \cup S_2'(s), \\
S_1'(s) &= \{(i, j, k), 1 \le i \le j < k \le s\}, \\
S_2'(s) &= \{(i, j, k), 1 \le k < j \le i \le s\}, \\
n &= 2s, \quad \bar{i} = i+s, \quad \bar{j} = j+s, \quad \bar{k} = k+s, \\
w_{pq} &= \begin{cases} 1 & \text{if } p=q, \\ 0 & \text{if } p \ne q, \end{cases}
\end{aligned}
$$

and

$$\Sigma^* \quad = \sum_{k=1}^{s} \quad \text{such that if } i=j \text{ then } k \neq i.$$

It can be seen that the number of terms in T_0 is $8(s^3-s)/3$, and each of T_1, T_2, T_3 has $8s^2$ terms. Therefore the complexity of the algorithm is

$$8(s^3-s)/3 + 24s^2 = (n^3-4n)/3 + 6n^2; \quad \text{this is still of } O(n^3).$$

The overall reduction in the number of M comes from the low coefficients of the dominant term in the complexity function. In Table 2.4.1 complexity of the Pan algorithm is shown for different matrix sizes, n. Table 2.4.2 gives comparative figures on the number of M required by the different methods.

Table 2.4.1 Complexity function growth of the Pan algorithm for different matrix sizes

n	Multiplications (M)	$\log_n M$
16	2 880	2.873
32	17 024	2.811 1
64	111 872	2.795 25
66	121 880	2.795 17
68	132 464	2.795 13
70	143 640	2.795 12
72	155 424	2.795 15
74	167 832	2.795 2
76	180 880	2.795 3
78	194 584	2.795 4
80	208 960	2.795 5
128	797 184	2.801

Table 2.4.2 The number of multiplications required to calculate the product of two $n \times n$ matrices using different methods

n	Traditional	Strassen's	Pan's
2	8	7	24
4	64	49	112
6	216	189	280
8	512	343	544
16	4 096	2 401	2 880
32	32 768	16 807	17 024
64	262 144	117 649	111 872
128	2 097 152	823 723	797 184

From these tables we see that the Pan algorithm is optimal at $n = 70$ giving the complexity function

$$T = O(n^\alpha). \quad \text{where } \alpha = 2.79512.$$

This compares favourably with Strassen's method for which $\alpha = 2.8074$.

Further improvements on the decomposition of the trilinear form lead to an algorithm of time complexity of $O(n^\alpha)$, $\alpha = 2.780\,142$ which is achieved for $n = 48$. The Pan results have recently been improved and the best result known is $O(n^{2.496})$ and is due to Coppersmith and Winograd (1982).

There is still room for further improvement of the results in the area, since the best-known lower bound on the evaluation of a set of bilinear forms is $2n^2 - 1$ (Brockett and Dobkin, 1978).

2.5 Summary

We have seen in this chapter that efficient algorithms for solving the problems of arithmetic complexity are frequently based on a technique known as *recursion*. Recursion is an important algorithm design technique. It is a method of solving a problem by dividing it into a small number of smaller subproblems of the same type as the original problem. The subproblems are divided in the same way. Eventually the subproblems become small enough to be solved directly. The solution to the smaller subproblems are then combined to give solutions to the bigger subproblems, until the solution to the original problem is computed.

It is very important for the applicability of the recursive approach that the original problem can be decomposed into only a small number of subproblems, otherwise a recursive algorithm may become extremely expensive.

Another useful rule for recursively partitioning the problem is to create the subproblems of approximately equal size; being able to do this often turns into a crucial factor in obtaining a good performance algorithm.

Using recursion one is able to state algorithms much more simply than would be possible without recursion: recursion simplifies the structure of many programs. Most of the modern programming languages allow recursive algorithms directly on a computer. The drawback here is that the way in which recursion is normally implemented within a language makes the run time very expensive, since procedure calls are much more costly than assignment statements.

3

Solving Recurrence Relations

In Chapter 2 we have seen several efficient algorithms whose derivation is based on recursion. In a more general setting the recursion as an algorithm design technique is known as 'divide-and-conquer' strategy.

3.1 Divide-and-conquer Approach

A typical application of divide and conquer has the following recursive structure.

> Given is problem *P*.
> **if**   *P* is divisible into smaller problems **then**
>   Divide *P* into two or more parts: $P_1, P_2, \ldots, P_k$;
>   Solve P_1;
>   Solve P_2;
>   . . .
>   Solve P_k;
>   Combine the *k* partial solutions into a solution for *P*
> **else** Solve *P* directly.

Whenever combining solutions of smaller subproblems of a problem is significantly simpler than solving the problem directly, the divide-and-conquer approach produces rather efficient algorithms, and therefore the method is widely used in the development of algorithms in many different computation areas.

In order to analyse the time or space performance of a recursive algorithm we can usually write down an expression describing the time or space function of the whole problem in terms of the time or space function for a smaller problem instance, e.g. in the problem of the multiplication of two $n \times n$ matrices, in order to estimate the number of pairwise multiplications of the matrix elements, we assumed *n* to be a power of 2, $n = 2^{s+1}$, and derived the expressions

$$
\begin{aligned}
M(0) &= 1, \\
M(s+1) &= 7M(s),
\end{aligned}
\tag{3.1.1}
$$

which tells us that the number $M(x)$ for the two matrices of size $n = 2^{s+1}$ each is expressed as seven times the number $M(x)$ for the two matrices of size $n = 2^s$ each, and initially, for the two matrices of size $n = 2^0 = 1$, this number is equal to 1.

The formulae of this kind are called *recurrence relations* (or simply *recurrences*). The first equation or equations of the system, which gives an explicit value of the function for a specific value of the argument is called the *boundary condition*.

Another example of a recurrence relation used in the problem of matrix multiplication is the equation which relates the number of additions of the matrix elements, i.e.

$$
\begin{aligned}
A(0) &= 0, \\
A(s+1) &= 7A(s) + 18(4^s),
\end{aligned}
\tag{3.1.2}
$$

where $n = 2^{s+1}$.

Recurrence is an elegant way of expressing an equation for the time characteristics of a recursive algorithm. The problem is then to deduce an explicit formula for the unknown function, $M(s)$. For equations (3.1.1) and (3.1.2) the solution is easily obtained by direct expansion of the equation, applying the recurrence itself over and over again until we arrive at the simplest case of the $M(s)$ for which an explicit value is given:

$$
\begin{aligned}
M(s) &= 7M(s-1) \\
&= 7(7M(s-2)) = 7^2 M(s-2) \\
&= 7^2(7M(s-3)) = 7^3 M(s-3) \\
&= \ldots \\
&= 7^s M(0) = 7^s, \quad \text{since } M(0) = 1.
\end{aligned}
$$

The formula $M(s) = 7^s$ is known as a 'closed form' for $M(s)$, and the technique for solving recurrences, as *the expansion of recurrences*.

3.2 Technique of Recurrence Expansion

In the problem of multiplication of two n-digit integers, the time cost of the computation, $T(n)$, is estimated by the number of multiplications of two one-digit numbers (the latter referred to as the elementary multiplication). The number of elementary multiplications required by the method (2.2.2) is given by the recurrence

$$
\begin{aligned}
T(1) &= 1 + c \leqslant c_1 \quad \text{(such } c_1 \text{ can always be found)}, \\
T(n) &= 4T(n/2) + c,
\end{aligned}
\tag{3.2.1}
$$

where c and c_1 are constants and c represents the overhead cost of the computation.

Similarly, the number of elementary multiplications needed by the method (2.2.3) is

$$
\begin{aligned}
T(1) &= 1 + b \leqslant b_1, \\
T(n) &= 3T(n/2) + b,
\end{aligned}
\tag{3.2.2}
$$

where b and b_1 are constants and b represents the overhead cost of the computation.

Both recurrence systems can be solved by expansion. As happens in the problem, the crucial difference in the efficiency of the two ways to multiply two n-digit integers lies in the different factors associated with the first terms on the right-hand sides of equations (3.2.1) and (3.2.2). For equation (3.2.2) we have

$$
\begin{aligned}
T(n) &= 3T(n/2) + b \\
&= 3^2 T(n/2^2) + (1 + 3^1)b \\
&= 3^3 T(n/2^3) + (1 + 3^1 + 3^2)b \\
&= \ldots \\
&= 3^k T(1) + (1 + 3^1 + 3^2 + \ldots + 3^{k-1})b \\
&= 3^k b_1 + (3^k - 1)b/2 = O(3^k).
\end{aligned}
\tag{3.2.3}
$$

Assuming $n = 2^k$ we know $k = \log_2 n$, and further, since $3^{\log_2 n} = n^{\log_2 3}$, we have $T(n) = O(n^{\log_2 3}) = O(n^{1.59})$. Note that the solution of equation (3.2.1) gives

$$
T(n) = O(n^{\log_2 4}) = O(n^2),
$$

that is, no improvement on the standard method of finding the product of two n-digit integers.

3.3 Solution of Homogeneous Recurrences

The general form of the recurrence of the type considered in the above examples may be given as

$$
\begin{aligned}
S(1) &= 1, \\
S(n) &= aS(n/b) + d(n)
\end{aligned}
\tag{3.3.1}
$$

where n is the size of a problem which is divided into subproblems, each of size n/b, and the term $d(n)$ represents the time expended on 'piecing' together the solutions to the subproblems to make a solution for the whole problem. Equation (3.3.1) is one of the most common forms of the recurrences that arise in the complexity studies of algorithms. Strictly speaking, equation (3.3.1) applies only to n's that are an integer power of b; however, under a reasonable assumption that $S(n)$ is smooth, obtaining estimates for $S(n)$ for those values of n, gives enough information about the growth of $S(n)$ in general.

A repeated substitution for S on the right-hand side of equation (3.3.1), as was done in the earlier examples, gives

$$
\begin{aligned}
S(n) &= aS(n/b) + d(n) \\
&= a^2 S(n/b^2) + ad(n/b) + d(n) \\
&= a^3 S(n/b^3) + a^2 d(n/b^2) + ad(n/b) + d(n) \\
&= \ldots
\end{aligned}
$$

$$= a^k S(1) + \sum_{i=0}^{k-1} a^i d(n/b^i)$$

$$= a^k + \sum_{i=0}^{k-1} a^i d(n/b^i). \tag{3.3.2}$$

Noting that $k = \log_b n$, we get $a^k = a^{\log_b n} = n^{\log_b a}$, the expression of n to a constant power.

In general, the larger is a (the more subproblems are solved) the higher the exponent will be, and the higher is b (the smaller each subproblem) the lower will be the exponent.

In analogy with differential equations terminology, the first term, a^k, of $S(n)$ is called the *homogeneous solution* and represents an exact solution for the $S(n)$ when $d(n)$ is zero for all n. Since $d(n)$ reflects the cost of combining the solutions of the subproblems into the solution of the whole problem, its being equal to zero corresponds to the case when such combining can be done 'without any cost'.

To continue the analogy with differential equations, the second term in $S(n)$ is called *the particular solution*. This term depends on both the number of subproblems, a, to be solved and the cost of combining their solutions, $d(n)$. Thus, while searching for possible improvements in the design of an algorithm, the interrelation of two terms in $S(n)$ needs to be carefully studied. For example, if the homogeneous solution is larger than the $d(n)$, then finding a quicker way to combine subproblems will not improve the efficiency of the whole algorithm; instead the ways of decomposing a problem into fewer (decrease a) or smaller (decrease n/b) subproblems should be sought. This will lower the homogeneous solution and improve the overall performance time, $S(n)$.

The observations concerning the roles that each of the two terms in $S(n)$ play in forming an overall magnitude order of $S(n)$ were illuminated by Aho *et al.* (1974, 1983).

The particular solution in equation (3.3.2) is very hard to evaluate in the general case, even if $d(n)$ is known. However, one special case of $d(n)$, a linear function in n, yields an explicit solution.

Suppose in equation (3.3.1) $d(n) = cn$, where c is a constant, then

$$\sum_{i=0}^{k-1} a^i d(n/b^i) = \sum_{i=0}^{k-1} a^i (cn/b^i) = cn \sum_{i=0}^{k-1} (a/b)^i, \tag{3.3.3}$$

and there are three cases to consider.
1. If $a < b$ then $a/b < 1$, giving

$$cn \sum_{i=0}^{k-1} (a/b)^i = O(n)$$

and

$$T(n) = n^{\log_b a} + O(n) = O(n^{\log_b a}). \tag{3.3.4}$$

2. If $a = b$ then

$$cn \sum_{i=0}^{k-1} (1)^i = cn \log_b n$$

and

$$T(n) = n^{\log_b a} + cn \log_b n = O(n \log_b n). \tag{3.3.5}$$

3. If $a > b$ then $a/b > 1$, giving

$$cn \sum_{i=0}^{k-1} (a/b)^i = cn \frac{(a/b)^k - 1}{a/b - 1} = O(n^{\log_b a})$$

and

$$T(n) = n^{\log_b a} + O(n^{\log_b a}) = O(n^{\log_b a}). \tag{3.3.6}$$

An analysis of the solution similar to the above can be carried out for a more general case of $d(n)$, when $d(n)$ is a power of n, higher than unity, i.e. $d(n) = n^\alpha$. In this case in equation (3.3.2) noting that $n = b^k$ we have $d(n/b^i) = d(b^{k-i}) = (d(b))^{k-i}$ and the particular solution of equation (3.3.2) becomes

$$\sum_{i=0}^{k-1} a^i (d(b))^{k-i} = d(b)^k \sum_{i=0}^{k-1} (a/d(b))^i$$

$$= d(b)^k \frac{(a/d(b))^k - 1}{a/d(b) - 1} = \frac{a^k - d(b)^k}{a/d(b) - 1} \tag{3.3.7}$$

The three different cases of relations between the values of a and $d(b)$ can be considered in a way similar to the linear case.

The above discussion on solution methods for recurrences is adequate for the analysis of many recursive algorithms and data structures arising in practice. In the remainder of the chapter a more systematic outline is given of the general methods for solving recurrences. The survey is presented in the spirit of the paper by Lueker (1980).

3.4 Homogeneous Equations

A homogeneous linear recurrence with constant coefficients has the form

$$p_0 s_n + p_1 s_{n-1} + \ldots + p_k s_{n-k} = 0, \tag{3.4.1}$$

where each p_i is a constant and s_i is an unknown function. The term 'linear' reflects the fact that a value is specified for some linear combination of the s_i, and the term 'homogeneous' refers to the fact that the linear combination is set equal to zero.

A general method for solving such recurrences uses the fact that the solution to the equation can be expressed as $s_n = r^n$, where r is some constant. Substituting the expression into the recurrence we get

$$\sum_{i=0}^{k} p_i r^{n-i} = 0, \tag{3.4.2}$$

which is satisfied provided

$$\sum_{i=0}^{k} p_i r^{k-i} = 0. \tag{3.4.3}$$

The kth degree polynomial (equation 3.4.3) in r is called the *characteristic equation* for the recurrence. We need to solve the polynomial for r to find its *characteristic roots*. A polynomial of degree k has k roots, which in general may be distinct or multiple. The following facts we shall state without proof.

(a) If there are k distinct roots, $r_1, r_2, \ldots, r_k$, then any linear combination of solutions to equation (3.4.2),

$$s_n = \sum_{i=0}^{k} c_i r_i^n, \quad c_i \text{ are arbitrary constants}, \tag{3.4.4}$$

is also a solution. Equation (3.4.4) is known as a general form of solution to equation (3.4.2). The boundary conditions associated with the recurrence determine the values of the constants in equation (3.4.4) and thus the solution we want.

As an example, consider the recurrence which defines the Fibonacci numbers:

$$\begin{aligned} F_1 &= 1, \quad F_2 = 2, \\ F_n &= F_{n-1} + F_{n-2}. \end{aligned} \tag{3.4.5}$$

The characteristic equation of (3.4.5) is

$$r^2 - r - 1 = 0$$

and the characteristic roots are

$$\frac{1 + \sqrt{5}}{2} \quad \text{and} \quad \frac{1 - \sqrt{5}}{2},$$

so that

$$F_n = k_1 \left(\frac{1 + \sqrt{5}}{2} \right)^n + k_2 \left(\frac{1 - \sqrt{5}}{2} \right)^n \tag{3.4.6}$$

for some constants k_1 and k_2.

Substituting the boundary conditions into equation (3.4.6) and evaluating the constants we eventually obtain

$$F_n = \frac{1}{\sqrt{5}} \left(\frac{1 + \sqrt{5}}{2}\right)^{n+1} - \frac{1}{\sqrt{5}} \left(\frac{1 - \sqrt{5}}{2}\right)^{n+1},$$

which with the use of the binomial theorem can be simplified to give

$$F_n = \frac{1}{2n} \left\{ \binom{n+1}{1} + 5\binom{n+1}{3} + 5^2\binom{n+1}{5} + \ldots \right\}.$$

Note that since the F_n's are integers the expression in the curly brackets is a number divisible by 2^n.

(b) If r is a root with multiplicity q then the equation has solutions

$$s_n = n^p r^n, \tag{3.4.7}$$

where p is any integer in $0 < p \le q-1$.

A more general expression for the solution function in this case is

$$s_n = P_{q-1}(n)r^n, \tag{3.4.8}$$

where $P_{q-1}(n)$ is a polynomial in n of degree less than q. Consider the following example of a recurrence:

$$\begin{aligned} h_0 &= 2, \qquad h_1 = 8, \\ h_n &- 4h_{n-1} + 4h_{n-2} = 0. \end{aligned} \tag{3.4.9}$$

This recurrence arises in the analysis of some characteristics of a perfect binary tree. The characteristic equation of (3.4.9) is

$$r^2 - 4r + 4 = (r - 2)^2 = 0,$$

and the general solution of the form

$$h_n = (c_1 + c_2 n)2^n.$$

Using the boundary conditions we obtain

$$h_n = (2n + 2)2^n. \tag{3.4.10}$$

(c) If a root of the characteristic equation is complex, it gives rise to a solution with periodic behaviour. Such solutions are most conveniently expressed in terms of the sine and cosine functions.

3.5 Non-homogeneous Equations

A recurrence similar to homogeneous but with a non-zero function of n on the right-hand side is called a non-homogeneous linear recurrence with constant coefficients, e.g.

$$h_0 = 2,$$
$$h_n - 2h_{n-1} = 2^{n+1}. \tag{3.5.1}$$

Any solution h_n which makes the left-hand side identically zero is called a homogeneous solution, and any solution h_n which satisfies the recurrence but not necessarily the boundary conditions is called a particular solution.

The general rule for solving a non-homogeneous equation is (i) to find the general homogeneous solution, (ii) to guess (propose the form of) a particular solution, (iii) to write down a combined solution of (i) and (ii), and to use the boundary conditions to solve for the constants.

We shall demonstrate the technique on the system (3.5.1). The general homogeneous solution is trivial and given as $h_n = k_1 2^n$; for the particular solution, after some effort we discover that $h_n = n2^{n+1}$ works, and so we get

$$h_n = k_1 2^n + n2^{n+1}.$$

Using the boundary condition we get $k_1 = 2$ and thus the solution

$$h_n = (2n + 2)2^n. \tag{3.5.2}$$

Guessing a particular solution when solving a non-homogeneous recurrence is an awkward feature of the method. Some kind of systematic procedure for producing such a solution would obviously be preferable. We shall outline one method for doing this which is applicable to many recurrences. First some useful notation is in order.

A sequence will be represented by writing down its nth element formula in curly brackets, e.g.

$$\{3^n\} \quad \text{means that for } n = 0, 1, 2, \ldots \text{ we have } 1, 3, 9, 27, 81, \ldots .$$

Let E be an operator which transforms a sequence by leaving out its first element so that, for example,

$$E\{3^n\} \quad \text{means } 3, 9, 27, 81, \ldots$$

and this fact will be abbreviated by writing

$$E\{3^n\} = \{3^{n+1}\}.$$

In general

$$E\{s_n\} = \{s_{n+1}\}. \tag{3.5.3}$$

Further, for any constant operator we can write

$$c\{s_n\} = \{cs_n\} , \tag{3.5.4}$$

and for the addition and multiplication of operators

$$(E_1 + E_2)\{s_n\} = E_1\{s_n\} + E_2\{s_n\} \tag{3.5.5}$$

and

$$(E_1 E_2)\{s_n\} = E_1(E_2\{s_n\}). \tag{3.5.6}$$

So defined, it can easily be verified that addition and multiplication of operators are commutative (i.e. $E_1 + E_2 = E_2 + E_1$ and $E_1E_2 = E_2E_1$) and associative (i.e. $(E_1 + E_2) + E_3 = E_1 + (E_2 + E_3)$ and $(E_1E_2)E_3 = E_1(E_2E_3)$); moreover, multiplication distributes over addition (i.e. $(E_1 + E_2)E_3 = E_1E_3 + E_2E_3$ and $E_1(E_2 + E_3) = E_1E_2 + E_1E_3$).

The following are two examples of applications of operators.

$$(E - 3) \{s_n\} = \{s_{n+1} - 3s_n\},$$
$$(2 + E^2) \{s_n\} = \{2s_n + s_{n+2}\}.$$

Observe that if $\Sigma_{i=0}^k p_i s_{n-i} = 0$ is a recurrence and $P_k(r) = \Sigma_{i=0}^k p_i r^i = 0$ is its characteristic equation then we may write the recurrence as

$$P_k(E) \{s_n\} = \{0\}, \tag{3.5.7}$$

since

$$P_k(E)\{s_n\} = \left(\sum_{i=0}^k p_i E^i \right) \{s_n\}$$

$$= \sum_{i=0}^k p_i s_i + \sum_{i=0}^k p_i s_{i+1} + \ldots + \sum_{i=0}^k p_i s_{i+n} = \{0\}.$$

From this equation we conclude that sometimes a certain operator, when applied to a certain sequence, produces a sequence consisting entirely of zeros. Such an operator is termed an *annihilator* for the sequence. For example, $(E-3)$ is an annihilator for $\{3^n\}$ since

$$(E - 3) \{3^n\} = \{3^{n+1} - 3 \cdot 3^n\} = \{0\}.$$

A fairly general technique for solving non-homogeneous linear equations with constant coefficients is based on the use of an annihilator:

(a) Determine an annihilator for the right-hand side of the non-homogeneous equation;

(b) Apply the annihilator to both sides of the equation and solve the resulting homogeneous equation.

Here is the list of some sequences and corresponding annihilators.

Sequence	Annihilator
$\{c\}$	$E - 1$
$\{$a polynomial in n of degree $k\}$	$(E - 1)^{k+1}$
$\{c^n\}$	$E - c$
$\{c^n$ times a polynomial in n of degree k$\}$	$(E - C)^{k+1}.$

The first three lines are special cases of the fourth line, so that if one wishes to prove correctness of the table, proving the last line will suffice. This can be done by first applying $(E - c)$ to $\{c^n P_d(n)\}$, where $P_d(n)$ is a polynomial of positive degree d, and obtaining $\{c^{n+1}Q_{d-1}(n)\}$, where $Q_{d-1}(n)$ is a polynomial of degree $d-1$, and then employing induction to establish the result.

Another useful fact is that if E_1 is an annihilator for $\{s_n\}$ and E_2 is an annihilator for $\{r_n\}$ then the product E_1E_2 is an annihilator for the sum $\{s_n + r_n\}$. For example, an annihilator for $\{n^23^n - 2\}$ is $(E - 3)^3(E - 1)$.

Three Examples on the Use of the Annihilator Technique

Three examples on the use of the annihilator technique are now exhibited. The first is equation (3.1.1). With our new notation we can write

$$(E - 4)\,\{A_{s+1} - 7A_s\} = (E - 4)\,\{18(4^s)\}, \qquad (3.5.8)$$

where $(E - 4)$ is the annihilator for the right-hand side of the recurrence. Equation (3.5.8) gives the recurrence

$$A_{s+1} - 11A_{s+1} + 28A_s = 0$$

and the characteristic equation

$$r^2 - 11r + 28 = 0.$$

Solving the latter we obtain the general solution in the form

$$A_s = k_1 7^s + k_2 4^s.$$

The constants are then determined from the boundary conditions, $A_0 = 0$ and $A_1 = 72$, and finally we have

$$A_s = 6(7^s - 4^s). \qquad (3.5.9)$$

The second example is equation (3.5.1), for which we have

$$(E - 2)\,\{h_n - 2h_{n-1}\} = (E - 2)\,\{2^{n+1}\}.$$

This leads to the characteristic equation

$$r^2 - 4r + 4 = 0,$$

after solving which we obtain the general solution

$$h_n = (k_1 + k_2 n)\,2^n.$$

The constants are determined using boundary conditions $h_0 = 2$ and $h_1 = 8$. Thus, the solution is

$$h_n = (2 + 2n)2^n.$$

As the third example, consider the recurrence

$$\begin{aligned} s_0 &= 0, \\ s_n - 2s_{n-1} &= 2^n - 1, \end{aligned} \qquad (3.5.10)$$

which arises in connection with the analysis of Mergesort. In this case we have

$$(E - 2)(E - 1)\,\{s_n - 2s_{n-1}\} = (E - 2)(E - 1)\,\{2^n - 1\}.$$

The resulting recurrence is

$$s_{n+2} - 5s_{n+1} + 8s_n - 4s_{n-1} = 0$$

and the characteristic equation

$$r^3 - 5r^2 + 8r - 4 = (r - 2)^2(r - 1) = 0.$$

Accordingly, the general solution is of the form

$$s_n = (k_1 + k_2 n)2^n + k_3,$$

which after solving for the constants and substituting their values into the s_n, becomes

$$s_n = (n - 1)2^n + 1. \tag{3.5.11}$$

3.6   Transformation Techniques

Sometimes a recurrence can be solved more easily after some transformation of the equation itself, its domain or its range. We now introduce some of the solution techniques based on such transformations.

Summing Factors

The following recurrence arises in the analysis of the average-case time performance of Quicksort.

$$t_1 = 0,$$

$$\frac{t_n}{n} = \frac{t_{n-1}}{n-1} + c\left(\frac{1}{n} + \frac{1}{n-1}\right), \tag{3.6.1}$$

where c is a constant, $n \geq 2$.

Now, observe what happens when we add up equation (3.6.1) for $n = 2$, $\ldots$, m.

$$\frac{t_2}{2} - \frac{t_1}{1} = c\left(\frac{1}{2} + \frac{1}{1}\right),$$

$$\frac{t_3}{3} - \frac{t_2}{2} = c\left(\frac{1}{3} + \frac{1}{2}\right),$$

$$\ldots$$

$$\frac{t_{m-1}}{m-1} - \frac{t_{m-2}}{m-2} = c\left(\frac{1}{m-1} + \frac{1}{m-2}\right),$$

$$\frac{t_m}{m} - \frac{t_{m-1}}{m-1} = c\left(\frac{1}{m} + \frac{1}{m-1}\right).$$

For $1 < n < m$, t_n occurs positively in one equation and negatively in the next. Thus when all these equations are added up, most of the terms will cancel out; in such cases the sum is said to *telescope*. The resulting equation is

$$\frac{t_m}{m} - \frac{t_1}{1} = c\left[\left(\frac{1}{2} + \ldots + \frac{1}{m}\right) + \left(\frac{1}{1} + \ldots + \frac{1}{m-1}\right)\right],$$

where the right-hand side is sum with the value

$$c(H_m - 1 + H_m - 1/m) = 2cH_m - c(m-1)/m,$$

and H_m denotes harmonic number $H_m = 1 + 1/2 + 1/3 + \ldots + 1/m$. Hence

$$t_m = m(2cH_m - c(m-1)/m) = 2cmH_m - c(m-1). \tag{3.6.2}$$

Finding the solution to equation (3.6.1) was straightforward. Sometimes a little more sophistication is required to make the sum of successive equations telescope. Suppose we have the recurrence

$$\begin{aligned} s_0 &= r_0, \\ p_n s_n - q_n s_{n-1} &= r_n, \qquad n \geq 1, \end{aligned} \tag{3.6.3}$$

where p_n, q_n, r_n are given.

It is obvious that unless $p_n = q_n$ (which is generally not the case), direct adding of two or more of these equations will not lead to cancellation of s_{n-1}. However, what we can do, is to multiply the successive equations of (3.6.3) by some factor which would force the s_{n-1} terms to cancel. We obtain

$$a_n p_n s_n - a_n q_n s_{n-1} = a_n r_n$$

and

$$a_{n-1} p_{n-1} s_{n-1} - a_{n-1} q_{n-1} s_{n-2} = a_{n-1} r_{n-1},$$

where the factor a_n is so far unspecified.

In order to make the s_{n-1} terms cancel, we require that

$$a_n q_n = a_{n-1} p_{n-1},$$

or

$$a_n = a_{n-1} p_{n-1}/q_n.$$

The latter can be enforced by letting

$$a_n = \prod_{i=0}^{n-1} p_i \bigg/ \prod_{i=0}^{n} q_i \tag{3.6.4}$$

(or any multiple of this constant). The telescopy then works for

$$a_n p_n s_n - a_n q_n s_{n-1} = a_n r_n$$

and for $n = 1, 2, \ldots, m$.

We shall now apply this technique to the recurrence (3.1.2):

$$A_0 = 0,$$
$$A_k - 7A_{k-1} = 18(4^{k-1}).$$

This is an example of equation (3.6.3) with $p_k = 1$, $q_k = 7$. By equation (3.6.4) we may choose $a_k = 7^{-k}$. Multiplying by a_k gives the new equation

$$7^{-k}A_k - 7^{-(k-1)}A_{k-1} = 18(4^{k-1}7^{-k}).$$

Summing this for k running from 1 to m gives

$$7^{-m}A_m - A_0 = (18/7) \sum_{i=0}^{m-1} (4/7)^i.$$

Thus

$$A_m = 18(7^{m-1}) \frac{1 - (4/7)^m}{1 - (4/7)} = 6(7^m - 4^m), \qquad (3.6.5)$$

which is naturally the same as in equation (3.5.9).

Knuth (1973) has a number of interesting applications of summing factors.

Range Transformation

Other useful transformations help the recurrence to appear in a more convenient form. Since a sequence can be thought of as a mapping from the integers into the reals, we can call a transformation on the values of the sequence a *range transformation*, and a transformation on the indices a *domain transformation*.

We will illustrate the technique of range transformation on the recurrence arising in a combinatorial problem. Suppose that n jobs have been assigned to n people. We wish to find in how many ways they can be reassigned the following day so that no person is given the same job as before. This problem is known as the derangements problem.

The recurrence that solves the problem of derangements is as follows:

$$s_1 = 0, \qquad s_2 = 1,$$
$$s_n = (n-1)s_{n-1} + (n-1)s_{n-2}, \qquad n \geqslant 3. \qquad (3.6.6)$$

This recurrence cannot be solved by any of the methods discussed so far, but if we let $s_n = n!t_n$ we may rewrite the recurrence as

$$t_1 = 0, \qquad t_2 = 1/2,$$
$$nt_n = (n-1)t_{n-1} + t_{n-2}, \qquad n \geqslant 3.$$

This new relation does not look more desirable than the original until we observe that it can be written as

$$n(t_n - t_{n-1}) = -(t_{n-1} - t_{n-2}),$$

which with a further transformation, $x_n = t_n - t_{n-1}$, becomes

$$x_2 = 1/2,$$
$$nx_n = -x_{n-1}.$$

This easily solved and we obtain

$$x_n = (-1)^n/n!, \qquad n \geqslant 2,$$

so that $t_n = x_n + t_{n-1}$, which by substitution gives

$$t_n = \sum_{i=2}^{n} (-1)^i/i!.$$

Finally

$$s_n = n! \sum_{i=2}^{n} (-1)^i/i! = n! \sum_{i=0}^{n} (-1)^i/i!. \qquad (3.6.7)$$

To demonstrate the technique of domain transformation we shall use one of the problems discussed earlier. In the Strassen method of matrix multi-plication, the product of two $n \times n$ matrices with $n = 2^{m+1}$, is computed by using seven multiplications of two matrices with halved size $n/2 = 2^m$, each. The recurrence is

$$M(2^{m+1}) = 7M(2^m),$$

where $M(2^k)$ is the number of pairwise multiplications of the matrices' entries required to find the product of two $k \times k$ matrices. The domain transforma-tion means the transformation on indices of a recurrence, and for the recurrence in hand, letting $s = \log_2(2^m) = m$, we can rewrite

$$M(s+1) = 7M(s),$$

which is undoubtedly a more manageable recurrence than the original equation.

Here is another example. In the method of 'ternary' search (where the search element is first tested at the $n/3$ place in the ordered sequence of elements, then, possibly, at the $2n/3$ place, and then, if the last test is unsuccessful, the size of the searched sequence is reduced to one-third of the original), one is led to the recurrence

$$C(1) = 1,$$
$$C(n) = C(n/3) + 2, \qquad n \geqslant 3, \qquad (3.6.8)$$

where n is required to be a power of 3, $n = 3^m$. Again, if we consider the recurrence

$$s_n = s_{n/3} + 2,$$

then we discover that none of the techniques we know are directly applicable to this equation. However, letting

$$n = 3^m \quad \text{and} \quad s_m = C(n) = C(3^m), \qquad m \geq 1,$$

we can write

$$s_0 = 1,$$
$$s_m = s_{m-1} + 2.$$

This is a recurrence which is easily solved by the methods already discussed, e.g.

$$s_1 - s_0 = 2,$$
$$s_2 - s_1 = 2,$$
$$\cdots$$
$$s_{k-1} - s_{k-2} = 2,$$
$$s_k - s_{k-1} = 2.$$

Adding together these equations we get

$$s_k - s_0 = 2k,$$

which is

$$s_k = 2k + 1, \qquad k = \log_3 n. \tag{3.6.9}$$

As the last example on the domain transformation, we solve the recurrence arising in the Schönhage–Strassen method of multiplying two n-bit integers for very large n (Schönhage and Strassen, 1971). The recurrence is

$$T(1) = 1,$$
$$T(n) = 2T(4n^{1/2}) + \log_2 n. \tag{3.6.10}$$

We wish to choose an index k so that the recurrence could be written as

$$s_k = 2s_{k-1} + (\text{some function of } k). \tag{3.6.11}$$

Letting n_k denote the value of n which corresponds to a given k and comparing the terms $2s_{k-1}$ and $2T(4n^{1/2})$ in equations (3.6.11) and (3.6.10) respectively, we must have

$$n_{k-1} = 4n_k^{1/2}.$$

Then, converting the non-linear recurrence into a more tractable form by the range transformation $m_k = \log_2 n_k$, we obtain

$$m_k = 2m_{k-1} - 4.$$

The solution of the latter is easily seen to be

$$m_k = 2^k + 4,$$

so

$$n_k = 2^{m_k} = 2^{2^k + 4}. \tag{3.6.12}$$

(Since we have not stated the boundary conditions, other solutions are also possible.)

If $s_k = T(n_k)$ then equation (3.6.10) becomes

$$s_k = 2s_{k-1} + 2^k + 4,$$

which is readily solved to yield

$$s_k = T(n_k) = k2^k + c2^k - 4,$$

where c is unspecified since the boundary conditions have not been given. So for any n appearing in the sequence $\{n_k\}$ it must be

$$
\begin{aligned}
T(n) &= O(k2^k) = O(\log_2(\log_2 n - 4)2^{\log_2(\log_2 n - 4)}) \\
&= O(\log_2(\log_2 n - 4)(\log_2 n - 4)) \\
&= O(\log_2 n \log_2 \log_2 n),
\end{aligned}
\tag{3.6.13}
$$

since from equation (3.6.12) $k = \log_2(\log_2 n_k - 4)$.

Using this solution one can show that the Schönhage–Strassen algorithm will multiply two n-bit integers in $O(n \log n \log \log n)$ time.

The Generating Functions Technique

A number of more ingenious methods for solving recurrences are based on the use of so-called generating functions. Like FTs or any other function transforms, generating functions transform a problem from one conceptual domain into another, frequently enabling the problem to be solved very elegantly. An interested reader is referred to the work of Knuth (1968), Liu (1968), Stanley (1978), Bender (1974), and Lueker (1980).

3.7   Guessing a Solution

Among many sophisticated methods for solving recurrences, an empirical approach whereby one simply tries to guess the solution is well respected.

In general, when one is faced with a recurrence to be solved and nothing whatsoever is known about its solution, it would be unwise to try to arrive at the solution by guessing what it should be; some systematic approach must be used whenever possible. Yet there are some situations in the complexity analysis where one has some information about, say, the form of the solution function, $f(n)$, up to some parameters, suitable values for which need to be deduced from the recurrence. In such cases, substitution of the proposed form of the solution into the recurrence can quickly yield to the desired solution.

We shall illustrate the approach on an example. The problem concerns finding the kth largest in a set of n unordered elements. Until 1973 it was believed that this problem can only be solved by first sorting the set in order and then picking out the required element. Since sorting a set of n elements in order is a problem of asymptotic complexity $O(n \log n)$ it was concluded that the same complexity is asymptotically optimal for the former problem. Then, Blum et al. (1973) showed that by a careful application of the divide-and-conquer strategy (hence the recursive solution) one can find the kth largest element in $O(n)$ time. Analysis of the method gives rise to the following recurrence:

$$T(n) \leq T(n/r) + T(3n/4) + cn, \qquad (3.7.1)$$

where r and c are constants and c is such that

$$T(n) \leq cn \quad \text{for} \quad n \leq n_0.$$

Since the method is anticipated to perform with linear complexity, the solution may be proposed in the form

$$f(n) = K(r)cn,$$

where $K(r)$ is some function of r.

Proof. The proof that the function of the form proposed satisfies the recurrence is given by means of the induction argument.

For the *induction base* we have

$$T(n) \leq cn \quad \text{for } n \leq n_0,$$

for the *induction hypothesis* we let

$$T(m) \leq K(r)cm \quad \text{for } n_0 < m < n,$$

and the *induction step* is then to show that

$$T(n) \leq K(r)cn \quad \text{for } m = n.$$

Now solve

$$\begin{aligned}
T(n) &\leq T(n/r) + T(3n/4) + cn \\
&\leq K(r)cn/r + K(r)3cn/4 + cn, \quad \text{for } r > 1, \\
&= K(r)(1/r + 3/4)cn + cn \leq K(r)cn,
\end{aligned}$$

provided $(1/r + 3/4) < 1$, that is, provided $r > 4$.   QED.

4

Complexity of Data-processing Problems

Data-processing problems include such problems as sorting, searching, and updating a sequence of elements.

A sorting problem assumes a collection of unordered elements which have to be sorted in a natural non-decreasing or non-increasing order. The nature of the problem implies that the elements of a collection are defined so that the notions of 'greater than', 'equal to' and 'less than' are meaningful within the collection.

A searching problem assumes an ordered set of elements and an element called the search element. One then searches through the ordered set to establish whether or not the search element is in the set.

One large group of algorithms for solving sorting and searching problems is based on the assumption that the complete set of elements is stored in the main memory of the computer and so at any particular moment of time one can access directly any element of the array.

If the problem assumes a very large set of elements then the storage aspects such as storing of the data in secondary memory, e.g. magnetic tape or disc, add new constraints on the design of the algorithms. When the elements of the set are complex aggregates of information in their own right, the set would normally be called a file of records.

In a typical file-sorting problem a file of records $R_1, R_2, \ldots, R_n$ is stored in either secondary or main memory. The index i of the record R_i indicates its location in memory. Each record R has a key, $K(R)$. The computational task is to rearrange the file in memory into a sequence $R_{i_1}, \ldots, R_{i_n}$ so that the keys are in non-decreasing order, $K(R_{i_1}) \leqslant K(R_{i_2}) \leqslant \ldots \leqslant K(R_{i_n})$.

The record as an item may be considerably larger than the key, and the problem is of actually rearranging the records and not the keys. This is a much harder problem than the sorting of relatively small sets of relatively simple elements, such as numbers.

4.1 Worst-case and Average-case Time Bounds

We shall illustrate on some examples the methods of complexity analysis which are used in derivation of time bounds.

Suppose we are given an array of n numbers, NUMBER $[1 \ldots n]$, and wish to find the largest value among them. A simple method is to scan the array

38

sequentially, comparing each element with the largest one found to that point.

Algorithm findmaximum

```
while n > 0 do
    index := 1
    max := NUMBER [index]
    while index < n do
            index := index + 1
            if max < NUMBER[index] then
                max := NUMBER[index]
            end if
    enddo
enddo
```

There are two basic operations involved in the algorithm. These are the comparison of the max value with the current element value of array NUMBER and the assignment of a new value to the max. The loop must be repeated $n-1$ times, and this sets lower bounds on the execution time of the algorithm; the assignment statement is always executed once, and so if the first element happens to be the largest one, the assignment is done only once; on the other hand, if the sequence is an increasing one, max is assigned a new value n times. Hence 1 and n are the extreme values. As for the average, there are $n!$ possible arrangements of the numbers.

In order to obtain the average number of assignment statement executions we could examine every one of $n!$ possible arrangements of n numbers, counting the assignment executions for each arrangement, then computing the total sum of the executions and dividing it by the number of arrangements, $n!$. This is too many to examine even for small values of n. Instead we can arrive at the result analytically.

We have noted that the initial assignment of the max value is always executed once, so the count of executions begins at 1. Assuming all permutations are equally likely, the probability that the second element is larger than the first is 1/2. The probability that the third element is larger than either of the first two elements is 1/3. Continuing in this way, we obtain the average of assignments equal to the sum

$$1 + 1/2 + 1/3 + \ldots + 1/n = H_n, \qquad (4.1.1)$$

which for large n is equal to

$$\log_e n + 0.577. \qquad (4.1.2)$$

Since the log n grows much slower than n itself, the time complexity of the algorithm is determined by the *do*-loop time and is thus proportional to n.

Some very efficient searching and sorting algorithms have been designed, and lower bounds proved on the complexity of these problems. For example it is known that the searching on a sequence of n numbers in a RAM machine requires about log n comparisons, and the sorting of n numbers about n log n comparisons.

Sorting Problem

The computational model for sorting four numbers, $K1$, $K2$, $K3$, $K4$ is shown in Fig. 4.1.1. It is a binary tree and each path from the root of the tree to its terminal node, denoted by $\square$, is meaningful, that is no redundant comparisons are allowed in the tree, so if $K1 < K2$ and $K1 > K3$ then it is 'known' that the ordered sequence is $K3 < K1 < K2$, without comparing $K2$ and $K3$.

If the root of the tree is on level 1 then at the lowest level, k, in the tree there are at most 2^k nodes. On the other hand there are $n!$ permutations of n numbers. This gives $2^k \leq n!$ or $k \leq \log_2 n!$ which for large n is approximately equal to n log n. Hence the lower bound on the number of comparisons required to sort n numbers is of order n log n. For some of the known comparison sorts, e.g. Heapsort, 2-way merge, and merge-insertion sort (Knuth, 1973, Section 5.3.1), this lower bound is the worst-case complexity. For other algorithms, e.g. Quicksort, $O(n \log n)$ is the average-case complexity, while the algorithm's worst-case complexity is of $O(n^2)$. We note that in the analysis it is assumed that all permutations on n input numbers are equally likely.

Worst-case complexity measures provide the algorithm's performance *guarantee*: the problem will always require no more time or space than the worst-case estimate. However, for some algorithms the worst-case bound may be overly pessimistic. For example, if a problem is such that its solution requires 10^{12} operations 1 per cent of the time but only 1 000 operations 99 per cent of the time by the worst-case bound judgement, the problem is very slow. Yet in many problems one is concerned with the average or typical efficiency of solving the problem instance. In particular, for sorting and searching problems, because of their frequent use in various applications, average-case analysis is almost always more realistic than worst-case analysis.

File-sorting Problem

The file-sorting problem where a file of records $R_1, \ldots, R_n$ is rearranged, poses additional constraints because the file usually resides in some sequential peripheral memory. Size limitations result in only a small number of records transferred at a time into fast memory for rearrangement. Yet it is possible to develop algorithms for the actual reordering of the files in time $O(n \log n)$. Floyd (1972) has shown that the redistribution of records in the above manner can be achieved by $k \log_2 k$ transfers into fast memory, given the following. A file distributed on a number of pages $P_1, \ldots, P_m$, each page containing k

41

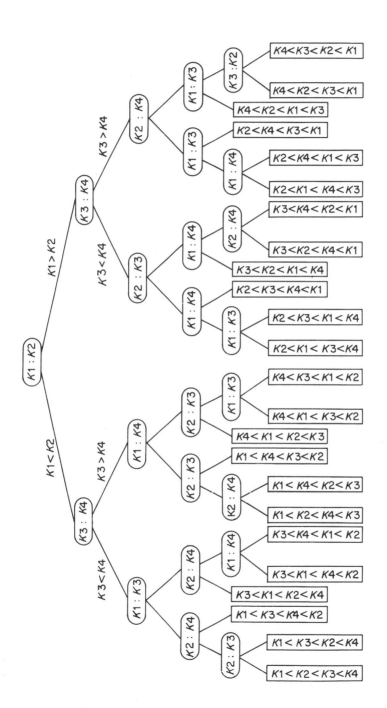

Figure 4.1.1 Binary comparison tree for sorting four distinct elements using the operation of comparison only

records, so that P_i contains the records $R_{il}, \ldots, R_{ik}$, where without loss of generality it is assumed that $m = k$, the task being to redistribute the records so that R_{ij} will go to page P_j for all $1 \leqslant i, j \leqslant k$, and the fast memory being large enough to allow reading in two pages, P_v and P_w, to redistribute their records and to read the pages out. This result, the best possible, was proved using a recursion analogous to that employed in the FFT.

A Search Algorithm

When a quick retrieval of information is important, a data structure known as the binary search tree is often used. The binary search tree has two notable properties: each node can have at most two children (subnodes), and the keys that identify the nodes are arranged so that at any node the smaller key is in the left subtree. In such a set-up searching for a particular key can be very efficient; a comparison at each node indicates whether to take the left branch, or the right, and so the number of remaining possibilities is halved at each node. The maximum efficiency is achieved when the tree is perfectly balanced, i.e. when every node has exactly two children. The average number of key comparisons for a successful search, i.e. when the key is found to be in the tree, is then equal to $\log_2 n$. In the worst case, where the tree has degenerated to just one child linked to each node, the average number of comparisons is one-half the number of nodes, $n/2$.

The two instances of the binary search tree are illustrated in Fig. 4.1.2. From the worst-case result it appears that one ought to take some care to keep the tree balanced as it grows. However, the balancing in itself requires considerable effort and so one might question whether the additional effort will pay off in faster searches. Surprisingly, in most cases it does not (Knuth, 1973).

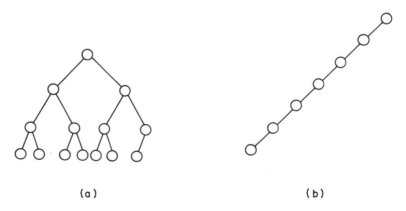

(a)   (b)

Figure 4.1.2 Binary search tree. (a) Perfectly balanced tree for 14 keys; (b) degenerated tree for 7 keys

Suppose n key values are read in random sequence and inserted into a tree, which is initially empty. The keys are placed in the left-to-right ordering, but no effort is made to balance the tree. A single procedure can be designed both to add a new key and to search for one that is already present.

Algorithm treesearch (treenode)

//The algorithm searches for key KK in a tree defined by treenode; if KK is not found, the algorithm inserts it.//

if treenode = nil   //the tree is empty//
then insert new node with key KK
else
    case KK **of**
       : $KK <$ treenode.key : *treesearch (left)*;
       : $KK >$ treenode.key : *treesearch (right)*;
       : $KK =$ treenode.key : found
    endcase
endif

The analysis of the average number of comparisons is simplified if we make use of the notion of the tree path length. The path length of the tree is the sum of the distances of all nodes from the root, and the distance of a node from the root is one less than its level; the root is on level 1.

Now consider the average path length in a tree constructed by such random insertion, noting that a node in the binary search tree represents the operation of comparison of two keys.

Assuming that the keys are the integers 1 to n and that all permutations of the keys are equally likely, let some key, i, arrive first and so become the root of the tree. When the rest of the keys have been inserted there will be $i-1$ nodes to the left of the root and $n-i$ nodes to the right. If i happens to be the midpoint of the range, the tree is perfectly balanced at this highest level; if i is equal to 1 or to n, the first branches of the tree are completely unbalanced. Continuing the argument recursively we note that if the second key to arrive is j and it is less than i, then when the tree is filled there will be $j-1$ nodes in the branch to the left of j and $i-j$ nodes in the branch to the right.

To calculate the average path length of the tree we forward the following argument (Wirth, 1984). At the root the path length is equal to 1 (for the root node itself) plus the length of a subtree with $i-1$ nodes plus the length of a subtree with $n-i$ nodes. These lengths are not known, but they can be calculated by applying the same procedure at the next level in the tree. Ultimately the end of each branch is reached, where every node has a path length of either 0 or 1. The recursive definition of the path length must be averaged for all possible values of i from 1 to n. We have

$$P_n(i) = [(i-1)(P_{i-1} + 1) + 1 + (n-i)(P_{n-i} + 1)]/n$$

and then

$$P_n = \frac{1}{n} \sum_{i=1}^{n} P_n(i) = 2(1 + 1/n)H_n - 3$$

$$= 2 \log_e n - 1.845, \quad \text{for large } n,$$

that is, the average path length of a tree constructed without concern for balancing is $O(\log n)$ and differs from the optimum only by a constant factor.

4.2 Data Structures

The algorithms discussed show that any algorithm, good or bad, depends on the data representation and data manipulation techniques within the computer memory. An algorithm may require one or more data structures, and the efficiency of the algorithm depends to a large extent upon good implementation of a given structure.

Two basic data structures on which all others are built are *arrays* and *linked structures*.

An *array* is a collection of memory locations numbered consecutively. The array is best suited for two operations: given the address of a memory location, (i) *store* a value in the memory location, overwriting the current value, or (ii) *retrieve* the current value from the memory location. One-dimensional blocks of data are normally represented by arrays.

A *linked structure* consists of a collection of records. Each record consists of a number of fields, each field having an identifying name. All records have identical structure. Fields are of two kinds: *data fields* and *reference fields*. Data is contained in data fields. *Pointers* to records are contained in reference fields. Two operations are associated with the linked structures: given a pointer to a record, (i) *store* a value into a field in the record or (ii) *retrieve* the current value from a field in the record.

While array addresses are integers capable of being manipulated by arithmetic operations, no operations are allowed on linked structure pointers except for storage retrieval and testing for equality. A linked structure can be implemented in one of the two ways, directly as a linked structure or as a collection of arrays, see Fig. 4.2.1.

Based upon arrays and linked structures, a variety of complex data structures can be implemented. The most commonly used among them are lists, unordered sets, ordered sets, graphs, trees, and mappings.

A *list* is a sequence of elements. The first element of a list is called *head*; the last element, *tail*. Lists are particularly flexible structures, they allow the following operations: (i) *scanning* the list to retrieve its elements in order; (ii) *adding* an element as the new head of the list, making the old head the second element; (iii) adding an element as the new tail; (iv) *deleting* and *retrieving* the

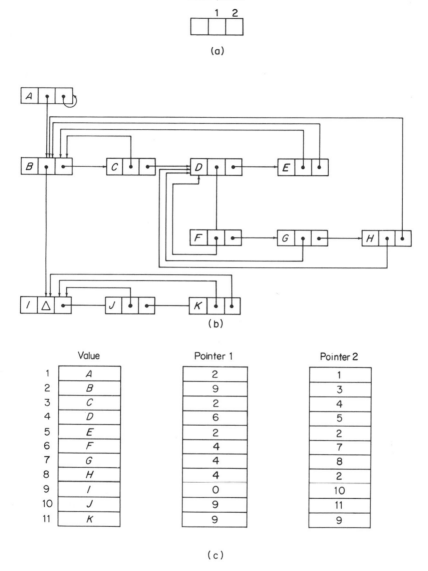

Figure 4.2.1 A linked structure and its implementation. (a) Record format; (b) linked structure; (c) representation of a linked structure by three arrays

head of a list, and deleting and retrieving the tail of a list; (v) two lists can be *concatenated* together, i.e. the head of the second list becoming the element following the tail of the first; (vi) one list can be *split* into sublists; (vii) an element can be inserted before or after an element whose location in the list is known; (viii) an element whose location in the list is known, can be deleted.

46

There are distinguished lists on which only a few of these operations are possible. These lists have special names.

A *stack* is a list with addition and deletion allowed only at the head. One particularly notable application of a stack is in the implementation of recursive procedures in programming languages.

A *queue* is a list with addition allowed only at the tail and deletion allowed only at the head.

A *deque* (double-ended queue) is a list on which addition or deletion is possible at either end. A deque can be implemented either as a *circular array*, i.e. addresses are computed modulo the size of the array, or as a *singly linked structure*, if deletion from the tail is not necessary, see Fig. 4.2.2(a), (b). The array representation uses no space for storing pointers, but requires that the amount of storage equals the maximum size of the list and be permanently allocated to the list. More difficult operations, like (v) to (viii) require a linked structure for their efficient implementation. For concatenation and for insertion after another element a singly linked structure is sufficient. *A doubly linked structure* is required for insertion or deletion before another element, see Fig. 4.2.2 (c). The list operations hardest to implement are inserting an element at the kth position in a list, retrieving the element at the kth position in a list, and deleting the element at the kth position in a list.

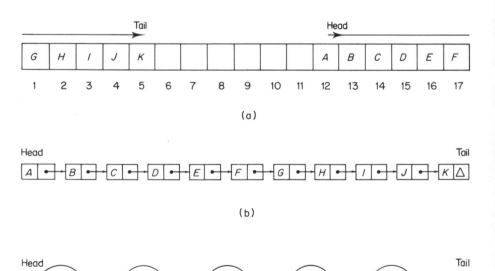

Figure 4.2.2 Representation of a deque containing *A, B, C, D, E, F, G, H, I, J, K.* (a) Array representation; (b) a singly linked structure; (c) a doubly linked structure

A *graph* is a set of vertices and a set of edges, each edge a pair of vertices. A graph can be represented by two-dimensional array **A**, called an *adjacency matrix*. The value of $A(i, j)$ is 1 if (i, j) is an edge of the graph, otherwise the value of $A(i, j)$ is 0. Another way to represent a graph is by an *adjacency structure*, which is an array of lists, one for each vertex. The list for vertex i contains vertex j if and only if (i, j) is an edge of the graph, see Fig. 4.2.3. If the graph is dense, i.e. most of the possible edges are present, the adjacency matrix representation saves space; it also takes constant time to test the presence of a given edge.

However, in 1973 Andreaa and Rosenberg conjectured, and in 1975 Rivest and Vuillemin proved, that testing any *non-trivial monotone* graph property requires the worst-case time proportional to n^2, where n is the number of vertices in the graph. (A graph property is *non-trivial* if for any n the property is true for some graph of n vertices and false for some other graph of n vertices. A graph property is *monotone* if adding edges to a graph does not change the property from true to false.) Alternatively a graph can be searched in $O(n+e)$ time, where e is the number of edges in the graph if an adjacency structure is used; thus representation by an adjacency structure is preferable for sparse graphs.

A *tree* is a graph without cycles. A tree imposes a hierarchical structure on a set of vertices. It can be defined recursively in the following way. A single node (vertex) by itself is a tree. This node is also the root of the tree. If n is a node and $T_1, T_2, \ldots, T_k$ are trees with roots $n_1, n_2, \ldots, n_k$ respectively, then a new tree is constructed by making n be the root and $T_1, T_2, \ldots, T_k$ the subtrees of the root. Nodes $n_1, n_2, \ldots, n_k$ are called the children of the parent node n. Since a tree is a graph it can be represented by an adjacency structure. A more compact way to represent a tree is to choose a root for the tree, compute the parent of each node with respect to this root, and store this information in an array, see Fig. 4.2.4. This representation is usable as long as the tree is to be explored from leaves (i.e. the nodes without children) to root, which is often the case in problems involving trees.

A *binary tree* is another useful and quite different notion of a tree. A binary tree can be either empty or a tree in which every node has either no children, a left child, a right child, or both a left and a right child. The fact that each child in a binary tree is designated as a left or right child, makes a binary tree different from a general tree. Both the arrays and linked structures are used for representation of binary trees.

For specific applications tree structures with special properties are used.

A *binary search tree*, which has already been discussed in the previous section, is a binary tree in which the nodes are labelled with elements of a given set, and all elements stored in the left subtree of any node N are all less than the element stored at N, while all elements stored in the right subtree of N are all greater than the element stored at N. This property, called the binary search tree property, holds for every node of a binary search tree, including the root. A binary search tree supports such operations as (i) *insert*

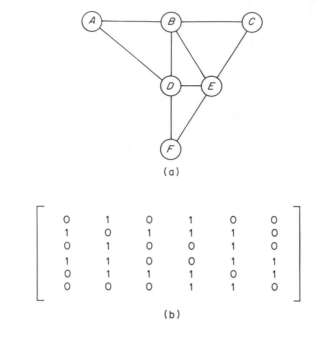

$$\begin{bmatrix} 0 & 1 & 0 & 1 & 0 & 0 \\ 1 & 0 & 1 & 1 & 1 & 0 \\ 0 & 1 & 0 & 0 & 1 & 0 \\ 1 & 1 & 0 & 0 & 1 & 1 \\ 0 & 1 & 1 & 1 & 0 & 1 \\ 0 & 0 & 0 & 1 & 1 & 0 \end{bmatrix}$$

(b)

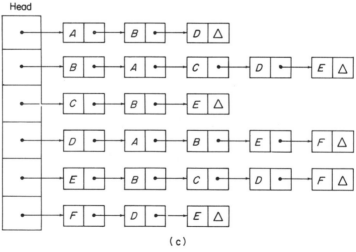

Figure 4.2.3   Representation of a graph. (a) Graph with six vertices and eight edges; (b) adjacency matrix; (c) adjacency structure

an element into the tree, (ii) *delete* an element from the tree, (iii) *test* whether an element is in the tree, and (iv) *find* the minimum value element of the set represented by the tree. Each of these operations can be carried out in $O(\log n)$ steps, on average, for a set of n elements.

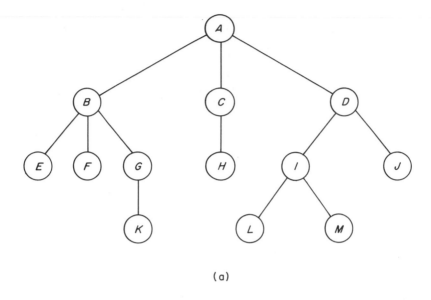

(a)

Parent

(b)

Figure 4.2.4 Representation of a tree. (a) Tree; (b) parent array for root A

An *AVL tree* (named after the inventors' initials) is a height-balanced binary search tree. In this tree the heights of two siblings are not permitted to differ by more than one level.

A *2–3 tree* is a tree with the following two properties. Each node has either none or two or three children and each path from the root to an external node has the same length.

It has been shown that AVL trees and 2–3 trees allow one to implement sorting and searching processes to run in $O(\log n)$ time (Aho *et al.*, 1974).

An *unordered set* is a collection of distinct elements with no imposed relationship. Each element of a set either is itself a set or is a primitive element called an atom. Main operations are (i) *adding* an element to a set, (ii) *deleting* an element from a set, (iii) *testing* whether an element is in a set.

There are two ways for a set implementation—by a singly linked list and by a *bit vector* (a bit one-dimensional array). In the case of a singly linked list representation, addition to an unordered set requires constant time, while testing and deletion require $O(n)$ time. For operations of comparison and sorting of the set values, the set can be represented by an AVL tree of a 2–3 tree in such a way that all three operations, addition, deletion, and testing require $O(\log n)$ time. If the number of possible elements is small, a set can be represented as bit vector. A *bit vector* is an array with one memory location for each possible element. A memory location has two possible values—true, indicating that the set contains the element, and false, indicating that it does not. All three operations require constant time using this representation (Aho *et al.*, 1974).

For large sets, the behaviour of a bit vector can be simulated by using a *hash table*.

A *hash table* consists of a moderate-size array and a *hashing function* which maps each possible element into an array address. An element, or pointer to it, is stored at (or near) the address specified by the hashing function. Since two or more elements may hash to the same address, some mechanism must be provided for resolving such collisions. With a hash table, addition, deletion, and testing require $O(n)$ time in the worst case but only constant time on average.

If two or more sets are manipulated simultaneously, additional operations such as forming a set which is the *union, intersection,* or *difference of two sets* are useful. Normally, for various representations, union, intersection, and difference require time proportional to the sum of the sizes of the sets.

An *ordered set* is a collection of numbered elements. Two important operations on an ordered set are (i) *sorting* the elements in increasing order, and (ii) *selecting* the element with the kth largest value. A variety of ways exist to sort n elements in $O(n \log n)$ time. Selecting the kth largest element requires $O(n)$ time (Blum *et al.*, 1973; Schönhage *et al.*, 1975).

A *priority queue* is an ordered set on which the following operations are allowed: (i) *adding* an element to the queue; (ii) *retrieving* the minimum-value element in the queue; (iii) *deleting* an element whose location is known from the queue. By using *binomial trees* (Vuillemin, 1978; Brown, 1977), *leftist trees* (Knuth, 1973), or 2–3 trees, the priority queue operations can be implemented so that they require $O(\log n)$ time. These implementations also allow one to combine two queues into a larger queue, destroying the smaller queue, in $O(\log n)$ time.

A *mapping* is a function, which associates the elements of one set with the elements of another set. The first set is called the *domain set*, and the second set the *range set*. The domain set or both sets can be finite or infinite. Certain mappings such as

$$ABS(x) = \begin{cases} x, & \text{if } x > 0, \\ -x, & \text{if } x < 0, \end{cases}$$

can be represented by an arithmetic expression or by some other means of calculating the mapping function values, while in many other cases the mapping function can only be described by a direct storing for each element of the domain set, the corresponding element of the range set (the mapping function values).

The operations performed on a mapping are: (i) *adding* an element to the domain set and stating its associated mapping function's value, i.e. the range value; (ii) *testing* whether a mapping function is defined for a given domain set; (iii) *assigning* a mapping function to a given domain set. Mappings are implemented as arrays, hash tables, or linked structures. An array representation, including a hash table, is used for mappings with a finite elementary type domain sets, whose range set is a sequence of elements with the defined first and last elements in the sequence. In general, any mapping can be represented by the list of pairs $(d_1, r_1), (d_2, r_2), \ldots, (d_k, r_k)$, where d_1, $d_2, \ldots, d_k$ are all the current elements of the domain and r_i is the value that the mapping associates with d_i, for $i = 1, 2, \ldots, k$. Such mappings can be implemented by a variety of linked structures.

A *trie* (the word derived from the middle letters of the word 'retrieval') is a special structure for representing sets whose elements are strings of objects of some type, e.g. character strings (words), strings of integers, etc. In a trie, each path from the root to a leaf corresponds to one element-string in the represented set. Tries are implemented as arrays or as linked lists.

4.3 Comments

Many large data-processing systems rely on few 'deep' algorithms; rather they are built up out of basic algorithms such as searching, sorting, and updating, which appear in many variations and combinations. On the other hand, the data structures tend to be exceedingly complex. The choice of the right data representation is often the key to successful data-processing programs.

5

Complexity and Difficult Combinatorial Problems

The field of combinatorial algorithms concerns the problems of performing computations, e.g. grouping, arrangement, ordering, selection, on discrete finite mathematical structures. Operations on matrices, the FFT algorithms, sorting of a file of elements, or searching for an element in a file are examples of combinatorial problems. We have seen that, as a rule, several algorithms are available for solution of these problems, and for some problems one can even point out the 'best' or an optimal algorithm. Typically, the time and space complexities of such an optimal algorithm are functions of the polynomial type, in the size of the problem. The algorithm is then called the polynomial time/space algorithm and its complexity, the polynomial time complexity.

In general, a polynomial complexity algorithm is defined as an algorithm whose complexity function is $O(p(n))$ for some polynomial function p and the problem size n. In practical terms, the polynomial time algorithm, when implemented on a real computer, requires a 'reasonable' running time and a 'reasonable' data storage, in order to arrive at the solution to the problem.

The theoretical studies of the complexity of computations have also indicated that it is unimportant what computer model is considered and what 'elementary computational steps' are available in the algorithm's repertoire. If an algorithm is found to be polynomial bounded when implemented on one type of computer, it will be polynomial bounded, though perhaps by a polynomial of different degree, when implemented on virtually any other computer.

An algorithm that cannot be bounded by a polynomial time function is termed an algorithm of exponential complexity. Table 5.1 shows some examples of functions of different order of complexity and the time required by algorithms with these complexity functions, to compute the problems solutions for different problem sizes, n.

A polynomial function grows much less rapidly than an exponential function, and there are even faster-growing functions such as a factorial function. In practical terms, the difference between the algorithms with polynomial and exponential complexities lies in a simple fact that when using a polynomial time algorithm one would normally obtain the computed solution, while with an exponential time algorithm one's whole lifetime would not be long enough to see the computed solution. The extreme cases of the

Table 5.1   Comparison of several complexity functions (We assume that the problem of size 1 takes 10^{-6} sec to compute.)

Time complexity function	Problem size		
	20	40	60
n	0.00002 sec	0.00004 sec	0.00006 sec
$n \log n$	0.00009 sec	0.0002 sec	0.0004 sec
n^3	0.008 sec	0.064 sec	0.216 sec
n^5	3.2 sec	1.7 min	13.0 min
$n^{\log n}$	0.4 sec	5.6 min	9 hr
2^n	1.0 sec	12.7 days	366 centuries
n^n	$3*10^{10}$ centuries		

two types of algorithms are the algorithms with the polynomial time of very high degree which render the algorithm impractical for applications, and the problem instances of very small sizes which can be solved by an exponential time algorithm. Both extreme cases are, however, of no practical importance.

5.1   Exponential Time Algorithms

The fundamental nature of the distinction between the algorithms of polynomial and exponential time complexities become transparent when considering the solution of large problem instances. Constant factors in the complexity function become less and less important as problem size increases; on large problems the asymptotic growth rate of the time bound dominates the constant factor.

Table 5.1.1 estimates the maximum size of problems solvable in a given amount of time. Increasing the amount of time or the speed of the machine by a large factor does not substantially increase the size of problems solvable unless the time bound grows more slowly than the exponential.

Table 5.1.1   Sizes of largest problem instances solvable in 1 hr on a computer with different computing speeds

Complexity function	With computer of certain speed	With computer of speed increased by a factor of 100	With computer of speed increased by a factor of 1000
n	n_1	$100n_1$	$1000n_1$
$n \log n$	n_2	$81n_2$	$739n_2$
n^3	n_3	$4.64n_3$	$10n_3$
n^5	n_4	$2.5n_4$	$3.98n_4$
$n^{\log n}$	n_5	$1.48n_5$	$1.77n_5$
2^n	n_6	$n_6+6.64$	$n_6+9.97$

Many combinatorial problems which obviously have some algorithms, seem to have no 'good' algorithms for their solution, in spite of the strenuous efforts to discover such algorithms. For instance, consider the maximum stable set problem: given a graph, find in it a maximum number of vertices, no two adjacent. A graph with n vertices has 2^n subsets of vertices. An algorithm for solving the problem is based on an exhaustive search. This gives the complexity function of $O(2^n)$, an exponential function, which even for moderate values of n become a formidable number. Since this number indicates the number of 'elementary computational steps' required by the algorithm, obtaining a solution for 'large' sizes of the maximum stable set problem quickly becomes infeasible. However, no one has yet discovered a substantially faster algorithm for this problem.

Other examples of the problems in the same category as the maximum stable set problem are:

(a) The travelling salesperson problem: find the shortest route for a salesman who must visit n cities.

(b) The k-clique problem: does an undirected graph on n nodes contain a complete subgraph on k nodes?

(c) The 0–1 integer programming problem: does a set of linear equations possess a 0–1 solution?

(d) The assignment problem: find a minimum sum subset of the elements in an $n \times n$ matrix, with exactly one element in each row and in each column.

(e) The problem of factoring a large number: find all the primes that divide the given large number evenly.

All these problems have been studied over a long time and the only algorithms for their solution that are known are of exponential computing time. Therefore, it has been conjectured that certain problems might not be at all solvable by 'reasonable' polynomial time algorithms.

It is important for practical applications to establish rigorously whether the problem is solvable by a polynomial time algorithm or it is not. This can be done by proving lower bounds on the complexity of the problem. Some important results have been obtained in proving lower bounds, but there still remain many open questions. For the problems known to be of polynomial time, e.g. sorting, matrix operations, and the like, the 'good' algorithms are generally made possible because of the gain of some deeper insight into the structure of a problem. Enormous efforts are applied at present to understand the crucial properties of 'difficult' combinatorial problems as well as similarly 'difficult' problems in other areas of applications of human activity, e.g. formal language theory and logic.

On the other hand, over the years the efforts in designing and development of algorithms brought about a number of design methods which are sufficiently general to be used in developing algorithms for many different classes of problems. Design methods such as recursion, optimization techniques, dynamic programming, graph search, branch-and-bound, backtracking, data-updating methods, and graph mapping often yield effective algorithms in solving large classes of problems. The next section gives a concise account of these methods.

5.2 Efficient Algorithm Design Techniques

An important part in the algorithm design process is working out various trade-offs to achieve an overall and as efficient as possible performance of the algorithm, for example, the time versus the storage, or the time versus the accuracy in a numerical problem.

Recursion

One of the oldest and widely useful algorithm design techniques is known as the divide-and-conquer approach.

Firstly, a number of specific design techniques are inherently recursive and so recursion is a natural way to describe algorithms obtained by these techniques. In its own right, as a divide-and-conquer method, recursion is an important algorithm design technique. It has already been illustrated extensively in the previous chapters, particularly in Chapter 2.

At this stage we would like to emphasize that when designing an algorithm, balancing the competing costs whenever possible is a useful rule in achieving an efficient design, and this is why the recursive algorithms which require division of the problem into subproblems of approximately equal size have, in general, a better overall performance. For instance, insertion sort where partitioning of the file of size n is done into $n-1$- and 1-sized subproblems gives the recurrence as

$$T(n) = T(n-1) + n$$

and hence is of complexity $O(n^2)$. On the other hand, Mergesort always partitions the n-sized file into nearly equal parts with recurrence

$$T(n) = T(\lfloor n/2 \rfloor) + T(\lceil n/2 \rceil)$$

and gives the complexity of $O(n \log n)$.

Recursion in general leads to exponential time algorithms because though only a small number of subproblems is created, some of them are resolved many times, giving exponential time complexity.

Optimization

A large class of problems, whatever their physical details may be, can be mathematically formulated as the problem where we have n inputs and are required to obtain a subset of the inputs that satisfies some constraints.

Any subset that satisfies these constraints is called a *feasible* solution. A feasible solution that either maximizes or minimizes a function defined on the problem (and called the *objective function*) is an optimal solution. We are required to find an optimal solution. There is normally an obvious way to determine a feasible solution but not necessarily an optimal solution.

Efficient algorithms available for the solution of such problems are often based on one of the two techniques, greed and augmentation.

The greedy method can easily be demonstrated on an example. Consider a postage stamp problem, where there is a set of stamps of six different denominations, 25p, 20p, 10p, 3p, 2p, and 1p. A particular letter we are arranging requires the total postage of 80p. We need to find the valid subset of the stamps. Almost without thinking we start with the largest denomination stamp, since all denominations are less than the total postage required, and build a subset element by element. We would get, say, 25p + 25p + 25p + 3p + 2p. This method of making up a postage fare is known as a greedy algorithm.

The greedy method suggests that at any individual stage a 'locally optimal' option, in some particular sense, is selected. However, it is easy to see that the postage fare solution in the above example can be 'improved' in terms of the smaller number of stamps yielding the needed postage total, e.g. four 20p stamps yield the same total as the five stamps chosen in the first instance; another four-stamp combination is two 25p, one 20p and one 10p. The example shows that not every greedy approach succeeds in producing the best result overall. It may produce a good result for a while, yet the overall result may be poor.

Still, there are many problems where greedy algorithms can be relied upon to produce 'good' solutions with high probability. For instance, if the problem is such that an exhaustive search is the only way to obtain an optimal solution, then the greedy method can be the only real, and wise, choice.

A classical example that illustrates this point is the travelling salesperson problem (TSP). As has been described, in this problem one searches for the shortest route for a salesperson who must visit *n* cities; every city must be visited and only once. With a view for its solution the problem can be formulated as a graph problem: we have an undirected graph with weights on the edges. It is reasonable for the TSP to assume that all edges exist, that is, that the graph is complete. A tour is a simple cycle that includes all the vertices. We wish to find a *tour* which minimizes the sum of these edge weights. In Fig. 5.2.1(a) an example is shown of the TSP, a graph with seven cities (vertices).

The only known algorithms that produce optimal solutions to this problem are of the 'try-all-possibilities' variety. They all have extremely expensive running times. On the other hand, the TSP has a number of practical applications and therefore efficient algorithms for the problem are essential.

A greedy algorithm for solving the problem considers the shortest edges first and accepts an edge into the solution subset if it, together with the edges already in subset, does not cause a vertex to have degree three or more, and does not form a cycle, unless the number of the edges in the subset equals the number of vertices in the problem; otherwise the edge is rejected. Collection of edges selected under these criteria will form a collection of unconnected paths, until the last step, when the single remaining path is closed to form a tour. An example of a greedy solution for a seven 'cities' graph is shown in Fig. 5.2.1(b).

The greedy algorithms are easy to implement and they are fast.

In situations where the greedy method does not work, sometimes an approach of *iterative improvements* can help. This general technique assumes a start with any solution to the constraints and then looks for a way to augment the weight of the solution by making local changes. The new solution is then improved again and in the same way. The process is continued until no improvement is possible. Such a locally optimal solution, under appropriate conditions, is also globally optimal. Often the augmentation method is a good heuristic. A very useful reference on how this approach can be applied to the TSP can be found in Lin (1965).

Dynamic Programming

This technique is a special kind of recursion applied to the problems for which there is no obvious way to divide a problem into a small number of subproblems whose solutions can be combined to solve the original problem. In such cases the problem is decomposed into as many subproblems as necessary and a track is kept of the generated subproblems so as to make sure that the same problem is never solved twice. This latter condition normally ensures a polynomial time algorithm as opposed to an exponential time algorithm of the direct recursion approach.

The most efficient way to achieve an overall control over the development of subproblems is to create a table of the solutions to all the subproblems one might ever have to solve. Sometimes one can discard the solutions for small subproblems as the computation proceeds and reuse the space for larger subproblems. The filling-in of the table of subproblems to obtain a solution for a given problem has been termed dynamic programming, hence the name of the technique.

After the table has been filled in, only certain solutions from the table are used to build up a sequence of solutions leading to an optimal solution to the original problem. This sequence of solutions, known as an optimal solution sequence, is arrived at by employing the property of optimality, which requires that for any intermediate state of the problem and the respective intermediate solution for this state, the solutions of the subsequent subproblems must constitute an optimal solution sequence with regard to the problem state resulting from the stated intermediate solution.

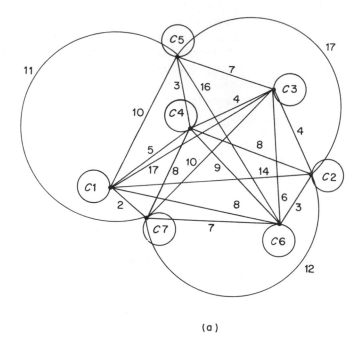

(a)

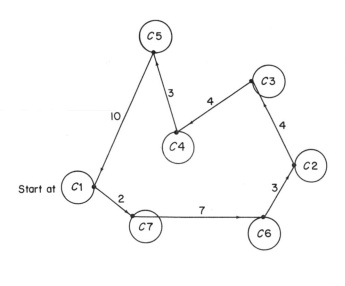

(b)

Figure 5.2.1 The TSP for an instance with seven cities. (a) Seven 'cities' and 'distances' between them; (b) a greedy method solution to (a); (c) start of a solution tree for the TSP instance in (a)

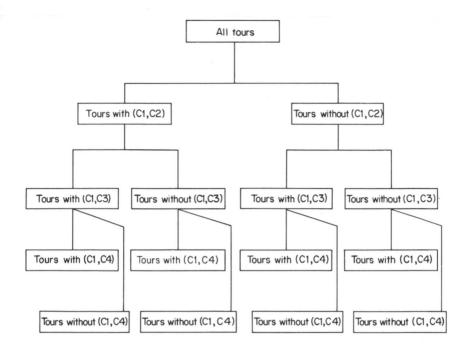

(c)

A dynamic programming algorithm may vary in form from problem to problem, but the filling-in of a table and the order in which this is done remain a common theme in all applications of the dynamic programming approach.

Graph searching

A fundamental problem concerning graphs is the path problem. In its simplest form it requires one to determine whether or not there exists a path in the given graph $G = (V, E)$ such that this path starts at vertex v and ends at u. The TSP is one example of the path problem. A more general form would be to determine for a given starting vertex $v \in V$ all vertices u such that there is a path from v to u. The problem is formulated as: find the connected components of the graph. This problem can be solved by starting at vertex v and systematically searching the graph G for vertices that can be reached from v.

The search is an examination of the edges of a graph using the following basic procedure.

Mark all edges and vertices of the graph 'new';
while 'new' vertex **do**
    Choose a 'new' vertex and mark it 'old';
    while 'new' edges lead away from the 'old' vertex **do**
        if the other endpoint of the 'new' edge is
        a 'new' vertex **then**
            Mark the edge 'old'
            Mark the endpoint-vertex 'old'
        endif
    enddo
enddo

If G is not connected the algorithm ensures a complete traversal of the graph. If G is connected, i.e. all vertices in the graph to be searched may be reached from the first start vertex, the search generates a *spanning tree* (i.e. a subset of edges of the graph that connects all nodes). The root of the spanning tree is the start vertex. The edges of the spanning tree are the edges which lead to new vertices when visited. The properties of the spanning tree depend upon the criteria used to select the starting vertex and the edges to explore. For some simple graph problems, such as finding connected components, any order of exploration is satisfactory. For more difficult graph problems, e.g. the TSP, the exploration order is crucial. There are two basic search methods.

In a *breadth-first* approach one starts at a vertex v. All unvisited vertices adjacent from v are visited next, i.e. the edge selected in the innermost *while*-loop is an edge out of vertex with candidate edges. Such a search partitions the vertices into levels depending upon their distance from the start vertex. A breadth-first search is implemented using a queue mechanism to store the old vertices.

If a *depth-first* search the exploration of a vertex is suspended as soon as a new vertex is reached. At this time the exploration of the new vertex u begins, i.e. the edge selected in the innermost *while*-loop is an edge out of the vertex u. When this new vertex has been explored, the exploration of v is continued. A depth-first search can be implemented as a recursive procedure or with an explicit stack to store the old vertices.

With respect to the two search methods the spanning trees are distinguished as the breadth-first and depth-first spanning trees. Figure 5.2.2 shows two such trees.

Both breadth-first and depth-first searches, if properly implemented using an adjacency structure to store the graph, require $O(n + e)$ time to explore an n-vertex, e-edge graph.

Other useful searches are *topological search* (Knuth, 1973), which labels the vertices of an n-vertex acyclic directed graph with integers from the set $\{1, \ldots, n\}$ such that the presence of the edge $\langle i, j \rangle$ in the graph implies that $i < j$, and *lexicographic search*, which searches for the lexicographically maximum spanning tree (Sethi, 1975; Rose *et al.*, 1976). Given two spanning trees which connect two vertices, v and u, of a weighted graph,

$$T_1 = \{\langle v_1, v_2 \rangle, \ldots, \langle v_{n-1}, v_n \rangle\}$$

and

$$T_2 = \{\langle u_1, u_2 \rangle, \ldots, \langle u_{m-1}, u_m \rangle\} ,$$

where $v_1 = u_1 = v$ and $v_n = u_m = u$. Here, T_1 is lexicographically greater than T_2 if there is some k such that

$$\text{weight}\ (\langle v_{i-1}, v_i \rangle) = \text{weight}\ (\langle u_{i-1}, u_i \rangle) \quad \text{for } 1 \leqslant i \leqslant k-1,$$

and

$$\text{weight}\ (\langle v_k, v_{k+1} \rangle) > \text{weight}\ (\langle u_k, u_{k+1} \rangle),$$

or else

$$\text{weight}\ (\langle v_{i-1}, v_i \rangle) = \text{weight}\ (\langle u_{i-1}, u_i \rangle) \quad \text{for } 1 < i \leqslant m.$$

A spanning tree connecting two vertices, v and u, which is not lexicographically less than any other tree connecting the same two vertices is said to be lexicographically maximum among the trees connecting v and u.

Two further techniques used for graph-solving problems are *shrinking* and *decomposition*.

Shrinking of a graph means that some part of it is replaced by a single vertex ('shrunk' to a single vertex)—often this part is a cycle or a union of cycles. The problem is then solved on the shrunken graph and from this the solution on the original graph is computed. This idea can be applied recursively, with the result that an overall faster solution to the original graph is achieved.

The decomposition approach is used when it is beneficial to partition the graph into several subgraphs, solve each subgraph, and combine solutions of the subgraphs to give the solution on the entire graph. In most instances where this technique is useful, the subgraphs are components (maximal subgraphs) satisfying some connectivity relation. In order to apply the technique, one must know an efficient way to determine the components. Several good algorithms exist for a variety of connectivity problems (Tarjan, 1972; Hopcroft and Tarjan, 1973a; Hopcroft and Tarjan, 1973b; Pacault, 1974; Tarjan, 1974; Tarjan, 1975).

Backtracking and Branch-and-Bound

It is a fact that the exhaustive search algorithms are not adequate for solving a difficult combinatorial problem of a practically interesting size. And still the concept of enumeration is at the heart of many of today's algorithms. Such algorithms render some usefulness because they employ various means of limiting the amount of enumeration that has to be done. They differ from one another only in how they go about it. Dynamic programming discussed above is one technique of this type. Another approach based on the same idea is known as branch-and-bound (BB) technique. It was first formulated by Little

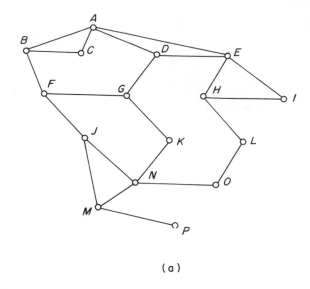

(a)

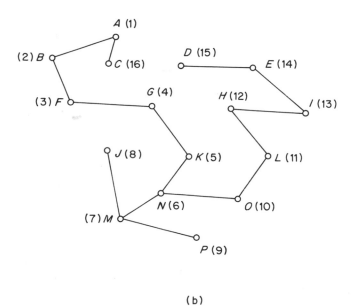

(b)

Figure 5.2.2 An example of a graph search. (a) An example of a connected undirected graph of 16 vertices; (b) a depth-first search of graph (a); (c) a breadth-first search of graph (b)

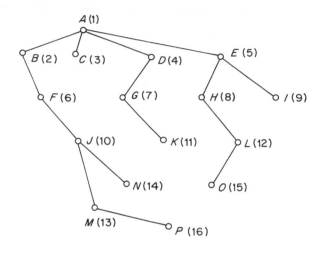

(c)

et al. (1963) in their landmark paper on an algorithm for solving the TSP. They introduced the name of the technique.

A typical BB algorithm contains three elements: partitioning into subproblems, selection of subproblems (branching) and bounding. Let us turn again to the TSP. For the TSP separation is done by dividing a given set of tours into two subsets; in one subset tours must pass over the link joining a certain pair of cities, and in the other subset no tour may use that link. As the enumeration proceeds the subsets decrease in size until it finally becomes possible to solve the subproblems. If this occurs before their number becomes enormous, the problem will be solved.

The most important possibility for limiting the number of subsets generated is the discovery of a sufficiently high lower bound for the subsets under investigation. If it can be shown, by some means, that all subsets of the current set can yield no lower value for the objective function than some value already obtained elsewhere, then that set need no longer be pursued.

When suitable strategies are used for the crucial choice of which subsets to explore first, so that high bounds are obtained soon and large areas of the problem are ruled out of searching, the technique is very powerful.

The systematic search of the feasible solution set for an optimum solution to the problem is organized as a tree. Many tree organizations may be possible. Figure 5.2.1(c) shows an example of the TSP and its tree organization for the BB search.

A backtracking algorithm uses the same ideas as the BB—it partitions the problem into a set of subproblems by systematically generating the nodes of a search tree with the solutions of the subproblems. Boundary functions are then used to reduce by as many as possible the number of nodes examined for their status as a feasible solution to the problem. Eventually the problem solution is obtained. The difference between the two methods lies in the way in which the tree of the subproblems solutions is generated.

Both methods begin with the root node. The remaining nodes, however, are generated by applying the depth-first approach in backtracking and the breadth-first approach BB. Boundary functions are used carefully enough so that at the conclusion of the process at least one answer node is always generated, or all answer nodes are generated if the problem requires us to find all solutions.

Data-updating methods

These are the methods which enable one to handle dynamic updating of data. The following can be mentioned among these techniques: the path compression method, the partition refinement technique, the linear arrangement technique.

The path compression solves the following problem. Given is a universe of elements, partitioned initially into singleton sets. Associated with each element is a value. The following operations may be requested to be carried out on the sets.

(a) *Union:* combine two sets into a single set, destroying the old sets.
(b) *Update:* modify the values of all elements in a given set in a consistent way.
(c) *Evaluate:* retrieve the value associated with a given element.

Partitioning refinement solves the following problem. Suppose the vertices of a graph are initially partitioned into several subsets. We wish to find the coarsest partition which is a refinement of the given partition and which is preserved under adjacency, in the sense that if two vertices v and w are contained in the same subset of the partition, then the sets

$$A(v) = \{x|(v, x) \text{ is an edge}\} \quad \text{and} \quad A(w) = \{x|(w, x) \text{ is an edge}\}$$

intersect exactly the same number of times with each subset of the partition.

The linear arrangement problem is as follows. Consider a set of n elements and a collection of subsets of the elements. Can the elements be arranged in a line so that each subset occurs contiguously? This problem arises in biochemistry and in archaeology.

5.3 Probabilistic Algorithms

When discussing performance of an algorithm in terms of the polynomial versus exponential complexity characteristics, it is customary to assume that the input data are distributed uniformly on the space of all instances of the problem.

The worst and the average times of the algorithm are then studied solely under these assumptions. As we have seen on a number of problems, the distinction between the worst-and the average-case behaviour of an algorithm is prompted by the fact that for certain problems, while an algorithm may require an inordinately long time to arrive at a solution for the least favourable instance of the problem, on the average the required time is appreciably shorter. From the practical point of view, when many instances of a problem have to be solved, the average behaviour is the more significant measurement of an algorithm. However, again the approach for analysing the average behaviour presupposes the existence of a known, e.g. uniform, probability distribution on the space of all instances of the problem in question.

Since the beginning of the 1970s an idea of using random numbers in solving difficult combinatorial problems has been conceived and implemented with exciting results. It was proposed that an assumption of a particular distribution on the space of instances of a problem is a drawback of the algorithm performance analysis, since for many important computational problems the relative frequency of the problem instances may be changing with time in an unpredictable and unmanageable way. The sample of instances actually appearing in a given application is often statistically biased in a manner not conforming to the assumptions made in the analysis of our algorithm.

To get away from the presupposed problem of the distribution of instances, a different approach and methodology for the study of algorithms was proposed: no assumption is made about the distribution of the instances of the problem to be solved. Instead randomization is incorporated into the algorithm itself.

For a problem instance I of size n the randomization is achieved by choosing at random an integer $1 \leqslant b \leqslant n$, or by choosing at random m integers, $b_1, b_2, \ldots, b_m$, all smaller than n. One then aims at constructing an algorithm involving a random step r so that for every instance I of the problem the average computation time will be brief. In this approach, randomness is not in the occurrence of that instances I, but is introduced into the algorithm itself.

The first such algorithm is attributed to Berlekamp (1970). It solves the problem of factoring a polynomial P of degree n over the field $GF(p)$ of p elements. The algorithm runs in time polynomial in $n \log p$, and, with probability at least one-half, it finds a correct prime factorization of P; otherwise it ends in failure. Since the algorithm can be repeated any number of times and the failure events are all independent, the algorithm in practice always factors in a feasible amount of time.

Since Berlekamp's, other efficient algorithms based on the same randomization principle have been developed. For instance, an algorithm for prime recognition of Solovay and Strassen (1977) accepts an input m and outputs either 'prime' or 'composite'. If m is in fact prime, the output is certainly 'prime', but if m is composite, with probability at most one-half the answer may also be 'prime'. The algorithm runs in time polynomial in the length of the input m.

The algorithm may be repeated any number of times on an input m with independent results. Thus if the answer is ever 'composite', it is known that m is composite. If the answer is consistently 'prime' after many, say, 100, runs it is a good indication that m is prime, since any composite m would give such results with a tiny probability, namely with the probability less than 2^{-100}.

We shall discuss in greater detail two probabilistic algorithms proposed by Rabin (1976). One of the algorithms determines whether a number is prime (this algorithm is different from that of Solovay–Strassen), and the other finds the nearest pair in a set of n points in R^k.

Rabin's Algorithm for Primality

Let n be a natural number. Determine whether n is prime. To do this:

Choose a random number b such that $1 \leqslant b \leqslant n$.
Test whether for b the following condition holds: either

$$b^{n-1} \neq \mod n, \tag{5.3.1}$$

or

$$\exists i[(n-1)/2^i = m \text{ is an integer}, \quad 1 < (b^m - 1, n) < n], \tag{5.3.2}$$

which reads 'there exists i such that $m = (n-1)/2^i$ is an integer and such that the greatest common divisor of $b^m - 1$ and n is greater than 1'.

The primality test condition (5.3.1)–(5.3.2) is due to Miller (1975) and states if condition (5.3.1) or (5.3.2) is true, n must be composite. Rabin calls this condition $W(b)$ and the integer b in this result a witness to the compositeness of n.

Repeat the test for m random numbers b, $1 \leqslant b_1, \ldots, b_m < n$. If $W(b_i)$ holds for any $1 \leqslant i \leqslant m$ then n must be composite. If for the sequence $r = (b_1, \ldots, b_m)$ of choices, $W(b_i)$ does not hold for any b_i, it is concluded that n is prime.

The analysis of the expected run time of the algorithm is faciliated by the result proven by Rabin:

There exists a constant c so that if n is composite then

$$(n-1)/2 \leqslant c \, (\{b \mid 1 \leqslant b < n, \ W(b) \text{ holds}\}), \tag{5.3.3}$$

that is, if n is composite then at least half of the $b < n$ are witnesses to the compositeness of n.

It follows from the theorem that for any fixed n, the probability of an error is smaller than $1/2^m$. Rabin's algorithm is found to be very fast. For example, the number $2^{400} - 593$ was identified as (probably) prime within a few minutes.

Comments on Probabilistic Algorithms

As can be seen on the Rabin algorithm, the basic requirement in a probabilistic primality algorithm is to find a test for compositeness for which the witnesses, in Rabin's terminology, are plentiful. If the test fails many times to produce a witness then one is provably confident that the number is prime. The same idea is used in the Solovay–Strassen primality algorithm.

In 1978 Rivest *et al.* published an important paper on public key cryptosystems. Their system requires the generation of large (100-digit) random primes. The testing of 100-digit numbers is done using the Solovay–Strassen method until the number is found that is provably prime in the sense outlined above. A more recent high-powered deterministic (as opposed to a probabilistic of Rabin and Solovay–Strassen) prime tester of Cohen and Lenstra (1982) can then be used to complete the test if it is important to know for certain that the number is prime.

The idea of a probabilistic test for 'witness' of a specified behaviour of the result can be considered in many other contexts. For example, let AL be a program or an algorithm and we wish to prove its correctness. One pragmatic method is to introduce data D, run AL on D, and check whether AL behaves as specified. The serious drawback is, of course, that, like, say, in the case of n being composite, the instances of D for which AL misbehaves may be rare. However, if for certain classes of program it is possible to enrich the set of 'witnesses' for incorrectness by using different tests, then a few randomly chosen tests will ensure a provably high probability of correctness.

Another observation of interest with respect to probabilistic algorithms is the fact that the computer-generated random number is, of course, a pseudo-random, i.e. fully deterministically generated, number. Nobody really knows how well pseudo-random numbers work for a given probabilistic algorithm. Experience so far shows that they seem to work well. But whether they will always work is not known. This brings about another line of thought: if the pseudo-random numbers always work well then true randomness would not help at all in improving the performance characteristics of the algorithms for solving a specified class of problems (Cook, 1983).

The Nearest-pair Problem

Rabin's primality algorithm is an example of a probabilistic algorithm where at certain junctures in the solution of a problem instance of size n by the algorithm, a random number b, $1 \leqslant b \leqslant n$ is chosen. If b_i is the ith number chosen then $r = (b_1, \ldots, b_m)$ is the total sequence of the random numbers produced. With the exception of the act of choosing the b_i, the algorithm proceeds completely deterministically. For simplicity, in such situations it is assumed that the process of random choice is such that for a given instance all sequences r are equally likely and the algorithm terminates for every choice r.

We now describe another model of a probabilistic algorithm where the possibility that the algorithm will occasionally produce an erroneous solution is permitted.

Let the algorithm be a probabilistic algorithm which for some instances of the problem and some choices r may produce an incorrect solution, and let c be a constant such that $c < \epsilon$. It is said that the algorithm solves the problem with confidence greater than $1 - \epsilon$ if for every problem instance the probability that the algorithm will produce an incorrect solution is smaller than ϵ. It is again assumed that the algorithm terminates for every choice of r.

The probabilistic algorithm of Rabin for finding the nearest pair in a set of n points is an example of a probabalistic algorithm which with very small relative frequency of the choice sequences r permits an occasional erroneous solution.

Let $x_1, \ldots, x_n$ be n points in k-dimensional space R^k. We wish to find (one of) the nearest pairs x_i, x_j for which

$$d(x_i, x_j) = \min d(x_p, x_q), \qquad 1 \leqslant p < q \leqslant n, \qquad (5.3.4)$$

where $d(x, y)$ is the usual distance function on R^k.

A brute-force method will yield the solution by evaluating all the $n(n-1)/2$ relevant mutual distances and their minimum. More sophisticated non-probabilistic algorithms require $O(n \log n)$ distance computations (Bentley and Shamos, 1976); they are recursive and as such involve considerable overhead in auxiliary computations.

A probabilistic algorithm of Rabin has the average number of distance computations of $O(n)$. We describe the algorithm assuming the Euclidean plane R^2, though the method is entirely general and the extension to R^k will be obvious.

The principal idea is to indentify clusters of the points $S = \{x_1, \ldots, x_n\}$ and to look for the nearest pair within clusters. If

$$S = S_1 \cup \ldots \cup S_k \qquad (5.3.5)$$

is a decomposition D of S and n_i is the cardinality of S_i, then the measure of D is defined as

$$N(D) = \sum_{i=1}^{k} \frac{n_i(n_i - 1)}{2}. \qquad (5.3.6)$$

If it is known that the nearest pair is within one of the S_i then it can be determined by computing

$$d(x_p, x_q), \quad x_p \in S_j, \quad x_q \in S_j, \quad p \neq q, \quad 1 \leq j \leq k. \quad (5.3.7)$$

This would involve $N(D)$ distance computations and $N(D)-1$ subsequent comparisons between the computed distances. A simple example of the 11-point nearest-pair problem is shown in Fig. 5.3.1.

To obtain the clusters of points a square lattice Γ of mesh size δ is drawn over the domain of the points, and the decomposition $S = S_1 \cup \ldots \cup S_k$ is associated with the lattice so that each S_i consists of all the points of S falling within one square of Γ. If $N(\Gamma)$ is the measure of this decomposition as defined by equation (5.3.6) then the aim of an efficient algorithm is to find a lattice for which $N(\Gamma) = O(n)$. This in turn may be achieved by choosing an initial lattice of mesh size δ and then repetitively doubling the mesh size until the desired result is obtained.

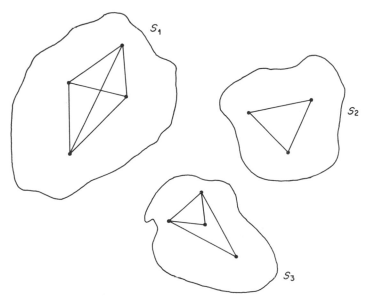

Figure 5.3.1   The nearest-pair problem for 11 points, where $S = S_1 \cup S_2 \cup S_3$ and $N(D) = 15$

The algorithm can be outlined as follows.

Select at random a sample of m points $S_1 = \{x_i, \ldots, x_i\}$ out of $S = \{x_1, \ldots, x_n\}$ and find $\delta = \delta(S_1)$.
This requires the average time of $O(n)$.

Construct a lattice Γ with mesh size $\delta(S_1)$.
(Rabin has proved that if $m = m(n)$ is appropriately chosen then with high probability $N(\Gamma) = O(n)$. His theorem actually states that there exists a constant c_2 so that if we choose at random $S_1 = \{x_i, \ldots, x_i\}$,

$m = n^{2/3}$, out of $S = \{x_1, \ldots, x_n\}$ and draw any square lattice Γ of mesh size $\delta(S_1)$, then the probability that $N(\Gamma) \leqslant c_2 n$ is greater than $1 - 2e^{-c_1 n^{1/6}}$, where c_1 is a constant of small size.)

Modify the lattice Γ by doubling the mesh size $\delta(S_1)$.
The four lattices $\Gamma_1, \ldots, \Gamma_4$ will follow, each with the mesh size $2\delta(S_1)$. One of the lemmas proved by Rabin ensures then that the nearest pair x_i, x_j in S falls within the same square of one of the Γ_i, $1 \leqslant i \leqslant 4$.

For each Γ_i find the induced decomposition

$$S = S_1^{(i)} \cup \ldots \cup S_{k_i}^{(i)}, \qquad 1 \leqslant i \leqslant 4, \tag{5.3.8}$$

where $S_j^{(i)}$ is a non-empty intersection of S with a square of Γ_i. Hence $k_i \leqslant 4n$. The average value of $N(\Gamma_i)$ is of $O(n)$.

Compute $d(x_p, x_q)$ for each $x_p, x_q \in S_j^{(i)}$.
The nearest pair is among these pairs so that

$$\delta(S) = \min d(x_p, x_q), \qquad x_p, x_q \in S_j^{(i)}, \tag{5.3.9}$$
$$1 \leqslant i \leqslant 4, \qquad 1 \leqslant j \leqslant k_i.$$

The computations of the distances involve certain list processing which is done by hashing in time $O(n)$, and further linear order time is expended on the comparisons of the distances.

The probability of obtaining an erroneous solution to the problem is smaller than $n(n - 1)e^{-cn^{1/6}}$, where c is a small constant.

Comments

A third model of a probabilistic algorithm conjectured by Rabin never produces an erroneous answer and terminates within a short average time but, with an infinitely small probability, may not terminate at all. Examples of a problem with such an algorithm have not yet been suggested.

6

Complexity and Theorem Proving by Machine

Ever since the advent of computers, one of the strongest and proudest ambitions of computer scientists was to endow the machine with some genuine powers of reasoning. Considerable efforts are continuously being undertaken in this direction. In particular, attempts were made to enable the computer to carry out logical and mathematical reasoning and this by proving theorems of pure logic or by deriving theorems of mathematical theories.

We shall consider one problem from the area of theorem proving by machine. Results obtained for this problem illuminate some important points of the complexity theory in general.

A few introductory definitions will be helpful. The simplest system which reflects something of mathematical reasoning is the propositional calculus, the formal language in which separate logical statements, which individually may be either true or false, are joined together by the lexical elements NOT, AND, OR, and IMPLIES. A logical system that has greater expressive power than the propositional calculus is the first-order predicate calculus. The basic statements in this language are formed from symbols representing *individual elements* and *predicates* (*properties*) and *functions* of them, and the compound statements are formed using the logical symbols of the propositional calculus together with ∀ ('for all') and ∃ ('there exists').

There is a precise notion of a proof of a statement of the predicate calculus, such that a statement is provable if and only if it is valid. Using the first-order predicate calculus it is possible to formulate a great deal of mathematics.

6.1 The First-order Integer Addition Problem

The system $N = \langle N, + \rangle$ is given which consists of the natural numbers $N = \{0, 1, \ldots \}$ and the operation + of addition.

The formal language L employed for discussing properties of the system (a so-called first-order predicate language) has

variables $x, y, z, \ldots$ ranging over natural numbers;
the operation symbol +, equality = ;
the logical connectives AND, OR, and IMPLIES;
the quantifiers ∀ and ∃, and parentheses.

A sentence s is a formula in which every variable is bound by a quantifier. For example, a sentence such as

$\exists x \ \forall y \ (x+y = y)$

is a formal transcription of 'there exists a number x so that for all numbers y, $x+y = y$'. This sentence is true in N (there is the null element in N).

Other examples are

$\forall x \forall y \ (x+y = y+x)$ is true, i.e. addition is commutative;

$\forall x \forall y \ (x = y+y)$ is false, i.e. not all numbers are even;

$\forall x \forall y \ [(\exists a(x+a=y) \text{ AND } \exists a(y+a = x) \text{ IMPLIES } x = y]$ is true, i.e. if $x \leqslant y$ and $y \leqslant x$ then $x = y$.

The set of true sentences in N is denoted by PA, Pressburger Arithmetic. In 1929 Pressburger showed that for the PA there exists an algorithm which on input of every given sentence s of the language L gives output YES or NO to the question whether this particular sentence is in PA. The PA is thus called 'decidable'.

6.2 Decidable and Undecidable Problems

The problem for which an algorithm exists is called decidable. An undecidable problem is a problem that cannot be solved by any algorithm. The concepts of decidability and undecidability are fundamental in mathematical logic, where it is a common task to decide whether a certain problem (or statement or theory) is true or false. Early investigations into the idea of effective computability were very much linked with the development of mathematical logic, as decidability was regarded as a basic question about any formalization of mathematics.

The undecidable problems establish formal framework in logic for questions that cannot, in principle, be resolved by computational means as we understand them.

In 1936 Turing demonstrated that certain problems are so hard that they are undecidable in the sense that no algorithm at all can be given for solving them. These classical undecidability results are the earliest *intractability* examples. Turing proved, for example, that it is impossible to specify any algorithm which, given an arbitrary computer program and an arbitrary input to that program, can decide whether or not the program will eventually halt when applied to that input. This is the famous Halting theorem.

In the same year Church showed that provability, and hence validity, in the predicate calculus is undecidable. This result was regarded by Hilbert as the most fundamental undecidability result for the whole of mathematics.

A variety of other problems have since been proved undecidable. One of the most significant results among these problems is due to Matijasevich (1970). It concerns the following problem.

Let $P(x_1, \ldots, x_n)$ be a polynomial in the variables $x_1, \ldots, x_n$ with integer coefficients. Then the equation

$$P(x_1, \ldots, x_n) = 0$$

for which integer solutions are sought is called a *diophantine* equation. Diophantine equations do not always have solutions. For instance, no integer value of x would satisfy the equation $x^3 - 2 = 0$. In 1900 Hilbert posed the problem, which since then is known as Hilbert's tenth problem (in Davis and Hensh, 1973). It asks whether there is an effective procedure that will determine whether any given diophantine equation has a solution.

Matijasevich has shown that there is no such procedure, that is, that Hilbert's tenth problem is undecidable.

6.3 Decidable Problems and their Algorithms

Proving the decidability of a problem is the first step in solving the problem: the PA is proved decidable, the propositional calculus theory is decidable, and many other problems are found decidable.

The next step is actually to construct an algorithm for obtaining the solution. As it turns out, there is a number of important problems that are decidable as far as the principle of the matter is concerned, but every algorithm which may be constructed for their solution takes such a vast amount of computing time that the problem remains 'practically undecidable' or 'practically uncomputable'. This fact first became apparent in the 1960s and early 1970s. The notion of an intractable problem was born then. Strictly speaking, an intractable problem is a problem which cannot be solved by a polynomial time algorithm. But this, in general, means that such a problem can either be solved by an exponential time algorithm or it can be undecidable. The intractable undecidable problems have a spirit of finality about them: once the problem is proved to be undecidable, that is the end of the matter. It is the intractable decidable problems that carry with them a real challenge in complexity theory.

The first examples of intractable decidable problems were obtained in the early 1960s, though they were 'artificial' problems, that is, specifically constructed to have the appropriate properties. Only in the early 1970s were the first 'natural intractable decidable' problems proved. Today there is a wealth of such problems proved, notably, in automata theory, formal language theory, and mathematical logic. The PA problem discussed earlier is one such problem.

Many attempts were made to devise an efficient algorithm to solve the PA. When such programs were run on computer, the computation terminated only on the simplest instances of the problem. Then in 1974 Fisher and Rabin showed that the reason for this is the fact that the inherent lower bound on complexity of algorithms for PA is a double exponential function of the input n. Their theorem states that

> There is a rational constant $c > 0$, such that if M is a Turing machine which accepts PA, then for infinitely many sentences s, M runs for at least $2^{2^{c|s|}}$ steps when started on input s.

Here s is the length of sentence s; it can, for example, be assigned as the number of occurrences of symbols $\forall$, $\exists$, $+$, $)$, etc. in a given sentence.

The fact that the algorithm defined as a Turing machine, requires this double exponentially growing time implies the same for all other definitions of algorithms including more realistic models of computers such as RAM.

From the Fisher–Rabin theorem it follows that the fact that the PA is decidable is of little use in designing practical solution algorithms: the explosive growth of the double exponential function suggests that any algorithm for PA will use hopelessly huge amounts of time on relatively short sentences thus rendering the problem practically unsolvable.

6.4 Diagonalization and Reduction

In a wide variety of situations, e.g. proving that there are more real numbers than integers (Cantor, 1874), proving that no method of proof could be both subject to mechanical verification and powerful enough to prove all theorems of elementary arithmetic (the Gödel incompleteness result, 1931), proving certain problems are not algorithmically solvable (Turing, 1936), proving lower bounds on the complexity of specific problems, etc. the validity or falseness of the statement can quickly be made transparent by assuming the opposite and deriving a contradiction from this assumption. For example, in Cantor's problem we can assume that the cardinalities of the two sets are equal and then show that this assumption is false. The two methods for securing such a contradiction are called *diagonalization* and *reduction*.

The diagonalization argument is based on ancient self-reference paradoxes and obtains a contradiction by a direct construction of an appropriately defined function based on its own index. Reduction, on the other hand, asserts contradiction indirectly by showing that if it is a wrongly assumed fact then it would contradict some already established result. Both techniques are powerful tools in the proofs of many results concerning the theoretical complexity of computations.

The basic idea of the diagonalization approach is well summarized by Cutland (1980).

> Suppose that ϕ_0, ϕ_1, ϕ_2, ... is an enumeration of objects of a certain kind, e.g. algorithms, functions, sets of natural numbers, etc. We can construct an object ϕ of the same kind that is different from every ϕ_n, using the following motto: 'Make ϕ and ϕ_n differ at n.'

The interpretation of the phrase 'differ at n' depends on the kind of object involved. Functions may differ at n whether they are defined, or in their values at n if defined there; with functions, there is usually freedom to construct ϕ so as to meet specific extra requirements, for instance, that ϕ be computable, or that its domain or range should differ from that of each ϕ_n.

Many proofs of undecidability rest on a diagonal construction. Tarjan (1978) has given an example on how the diagonal construction can be used to show that a set of *yes/no problems* is undecidable. There are many families of the so-called yes/no problems in computer science, and the computationally important question is to determine whether there exists an algorithm for solving all the problems in the family.

A family of yes/no problems is said to be decidable (solvable) if there is an algorithm which for every problem in the family will attain the correct 'yes' or 'no' answer. Conversely, if no such algorithm exists the family is said to be undecidable (unsolvable).

Now suppose we are interested in yes/no questions concerning the integers, such as 'Is n odd?' or 'Is n prime?'. Suppose further we have a listing A_1, A_2, ..., of all algorithms for answering such questions. Consider the set S of integers such that n is an element of S if and only if algorithm A_n answers 'no' (or does not answer at all) on input n. Then the question 'Is n an element of S?' is undecidable, since each algorithm in the list A_1, A_2, ... produces a wrong answer on at least one input (i.e. A_n is wrong on input n) and by Church's thesis (1932), which states that any algorithm, in the formal sense, can be expressed as a Turing machine, and any Turing machine expresses an algorithm, this list contains all possible algorithms.

Tarjan's is a typical example in that for the sets it is usually a question of membership: whether or not n is a member. In the example it is shown that there is no member of the set of the yes/no algorithms (which are supposed to be complete) which solves any problem of the family correctly.

An important part of this example is the reference to the Church thesis, sometimes the Church–Turing thesis. This thesis is one of the fundamental results in the computability theory. It is not a theorem (which would have to be proved), it has the status of a claim or belief which must be substantiated by evidence. The evidence in the support of the thesis is very impressive, so much so that many contemporary mathematicians accept it fully, and many leading results have been proved using the support of the Church thesis; these are known as 'proofs by Church's thesis' (Cutland, 1980).

Another example on the use of the diagonalization argument concerns the proof that there are functions that are not computable.

Theorem 6.4.1 There is a total (whose domain is the whole of $\mathbb{N}$) one-dimensional (unary) function that is not computable.

Proof. All we need is to construct a total function that is at the same time different from every function in the enumeration χ_0, χ_1, χ_2, ... of the unary computable functions. Let the new function be

$$f(n) = \begin{cases} \chi_n(n) + 1, & \text{if } \chi_n(n) \text{ is defined,} \\ 0, & \text{if } \chi_n(n) \text{ is undefined.} \end{cases}$$

For each n, f is different from χ_n at n:

if $\chi_n(n)$ is defined, then f differs from χ_n in that $f(n) \neq \chi_n(n)$;

if $\chi_n(n)$ is undefined, then f differs from χ_n in that $f(n)$ is defined.

Since f differs from every unary computable function χ_n, f does not appear in the enumeration of the unary computable functions and is thus not itself computable. From the definition it is clear that f is *total*. (A *total* function is a function defined over the whole infinite domain of its argument. See also Chapter 8.)   QED.

Note that we had a considerable freedom in choosing the value of $f(n)$; all we needed was to ensure that f is different from every $\chi_n(n)$. For instance,

$$h(n) = \begin{cases} \chi_n(n) + 3^n, & \text{if } \chi_n(n) \text{ is defined,} \\ 0, & \text{if } \chi_n(n) \text{ is undefined,} \end{cases}$$

is another example of a non-computable function.

6.5 Lower Bounds

A lower bound states a fact about all algorithms for solving the same problem. Since it is unrealistic to try and enumerate and analyse all these algorithms, deriving good lower bounds is a more difficult task than, say, devising a new algorithm. There are, of course, some problems for which the lower bound is an easily observable fact. For example, if we consider all algorithms for finding the largest of an unordered set of n elements, then we can see that in the process every element must be examined at least once and the lower bound is a linear function in the size of the problem, n. However, very few problems can be analysed in such a straightforward manner. Usually, proving lower bounds is a very hard and often frustrating activity; it is an activity which sets apart the complexity theory from the analysis of algorithms.

All important lower bounds on computation time and space are based on the technique of diagonalization. Turing and his contemporaries used the diagonal arguments to derive proofs about the algorithmically unsolvable functions. Prior to the 1960s the technique was used to define the hierarchies of computable 0–1 functions (Grzegorcyzk, 1953, in Cook, 1983). Then Rabin (1960) proved, using the diagonalization, that for any reasonable complexity measure, such as run time or space, sufficiently increasing the allowed time or space, etc. always allows more 0–1 functions to be computed. Later Rabin's result was extended in detail for time on multitape Turing machines by Stearns *et al.* (1965).

These early results gave lower bounds on the time and space to compute specific functions in the sense of directly relating the computation process to the computing machine action, e.g. counting the number of steps required to obtain the first digit of the output of machine M on input y. Then in 1972, Meyer and Stockmeyer gave proof of the exponential space and exponential time requirements by the equivalence problem for regular expressions with squaring. This was the first non-trivial bound for general models of computation on a 'natural' (i.e. not about computing machines) problem. Since the first results of Meyer and Stockmeyer there have been a large number of lower bounds on the complexity of decidable formal theories, among which the results on the time required to decide PA is one of the most interesting.

7

Classifying the Computational Complexity of Algorithms

The major notions in classifying the computational complexity of problems are the polynomial and exponential time algorithms, and in order to make the problems—very different by nature—comparable in terms of their complexity properties some further interpretation of these notions is due.

7.1 Terminology and Notation

We know that a polynomial time algorithm is an algorithm whose time complexity is of $O(p(n))$ for some polynomial function p in the size of the problem instance, n. In complexity measures which have been discussed so far each basic operation in the algorithm is counted as one step, then the measure gives the number of steps required. This measure is known as the *uniform cost measure*. Another way to measure an algorithm's performance is to change for an operation a time proportional to the number of bits needed to represent the operands. The measure based on this second approach is known as the *logarithmic cost measure*. In terms of the logarithmic cost measure, the size of the problem instance, n, is defined as the number of bits required to encode the problem instance.

For example, consider the TSP, where $G = (V, E)$ is a directed graph defining the instance of the problem. Let c_{ij} be the cost of edge $\langle i, j \rangle$ and let $|V| = v$. It is required to determine if a complete directed graph G has a tour of cost at most M. Assuming that all inputs are integer and that n is the length of the input to the algorithm, measured in binary representation, and further noting that, in general, a positive integer m has a length of $\log_2 m + 1$ bits when represented in binary, the input size of the TSP can be given as

$$n = \sum_{1 \le i, j \le v} (\lfloor \log_2 c_{ij} \rfloor + 1) + \lfloor \log_2 M \rfloor + \lfloor \log_2 v \rfloor + 2.$$

In such setting, the computational complexity, in similarity with the formal language theory, is concerned with the analysis of sets of 'strings of symbols' (rather than the input data sequences) or with 'languages' (rather than with the problems). Due to this similarity in the formulation of the aims of the computational complexity analysis and the formal language theory, many results of the formal language and automata theory are very useful in the analysis of theoretical computer science and computational complexity. As a result, the terminology such as a 'problem', a 'language', and a 'function' (as in the automata theory) are used interchangeably within the environment of computational complexity.

An algorithm whose time complexity cannot be bounded by a polynomial function is termed an 'exponential time' algorithm. The problem which is so 'hard' that no polynomial time algorithm can possibly solve it is termed an 'intractable' problem. An intractable problem, for some problem instances, may turn out to be so difficult that it is unsolvable, not in principle, but in practice.

The definition of 'intractable' provides a theoretical framework of considerable generality and power. Hartmanis and Hopcroft, (1971) have shown that the intractability of a problem is essentially independent of the particular encoding scheme and computer model used for determining time complexity. A number of problems are today seen to be intractable; the PA problem is an example of such a problem. Its lower bound on the time complexity is of $O(2^{2^{cn}})$, where n is the input length of the problem instance.

Another large group of problems is the set for which no efficient algorithms are known and the best available solutions require exponentially increasing time, yet no one has been able to prove that the problems do not have polynomial time solutions. These problems are termed as 'apparently intractable'. One of the major tasks of complexity theory is to develop means which would help to establish which natural problems are intractable and which are tractable, i.e. solvable in polynomial time.

One approach used in complexity theory to this end is proving upper and lower bounds on the time and space complexity of specific problems. These bounds are normally considered with respect to a worst-case performance measure of an algorithm. And though for some problems a worst-case bound may be too pessimistic—e.g. the Simplex method of linear programming (due to Dantzig, 1963) has an exponential worst-case time bound (Klee and Minty, 1972) but seems to run much faster than exponential on real-world problems—close upper and lower bounds if obtained, provide a realistic performance guarantee of the algorithm. There have been some important successes in proving lower bounds—see, for example, the developments concerning the linear programming problem in Section 7.4—but for a number of well-known problems there are still substantial gaps between known upper and lower bounds.

7.2 Non-deterministic Algorithms

Intractable and apparently intractable problems are much harder to understand in terms of their crucial properties, particularly in our usual environment, where the notion of algorithm presupposes the property that the result of every operation is uniquely defined. We call such algorithms deterministic and they agree with the way programs are executed on a computer.

Suppose now that the restriction on the outcome of every operation is removed. An algorithm is allowed to contain operations whose outcome is not uniquely defined but is limited to a specified set of possibilities. The machine executing such operations is allowed to choose any one of these outcomes subject to a certain termination condition. As an example, assume the following termination condition: whenever there is a set of choices that leads to a successful completion then one such set of choices is always made and the algorithm terminates successfully. This leads to the concept of a non-deterministic algorithm.

A non-deterministic algorithm terminates unsuccessfully if and only if there exists no set of choices leading to a success. A machine capable of executing a non-deterministic algorithm in this way is called a non-deterministic machine.

Before finding its way into the computational complexity field, non-determinism as a mathematical construct has been widely used in the studies of automata theory and formal languages, particularly when questions such as acceptance of languages by automata and their generation by grammars, and closure properties of classes of languages are studied.

Abstract automata are said to operate non-deterministically if the transition function (which specifies what to do at each step) is multi-valued—there is not a unique move to be performed at each step but rather a finite set of possible next moves. The automaton arbitrarily chooses (guesses) which move to perform and any choice is valid.

A non-deterministic algorithm may be interpreted as consisting of two stages, a guessing and a checking stage. The concept of a non-deterministic machine handles the guessing stage: it picks out at random an input string. It should be remembered that non-deterministic machines do not exist in practice. Their introduction provides theoretical (if intuitive) reasons to conclude that certain problems cannot be solved by 'fast' deterministic algorithms. At the checking stage the algorithm examines the input for a YES answer and outputs YES/NO result accordingly.

If a non-deterministic algorithm checks a given string of symbols for a YES answer in a polynomial time, it is termed a 'polynomial time non-deterministic' algorithm.

A deterministic interpretation of a non-deterministic algorithm can be made using the notion of a tree and allowing unbounded parallelism in computation (Floyd, 1967; Horowitz and Sahni, 1978). In the tree each node has branches corresponding to the possible next steps (decisions). All nodes on the same level are executed simultaneously. The first path in the tree to reach a successful completion terminates all other computations. If a path reaches a failure completion then only that path of the algorithm (tree) terminates.

While this interpretation may enable one better to understand non-deterministic algorithms, it is important to remember that the fundamental assumption about the non-deterministic machine is that it has the ability to select a 'correct' element from the set of allowable choices (if such an element exists) every time a choice is to be made. Whenever successful termination is possible, a non-deterministic machine makes a sequence of choices which is a shortest sequence leading to a successful termination.

The significance of the non-deterministic machine concept for the theory of computational complexity was first emphasized by Cook (1973), when he convincingly demonstrated the importance of the so-called *NP*-class problems (languages) that are accepted in polynomial time by non-deterministic Turing machines (or RAMs or programs written in some general-purpose language). A wide variety of important computational problems in combinatorial mathematics, mathematical programming, and logic followed which were shown to be in the *NP*-class (Karp, 1972; Cook, 1973; Aho *et al.* 1974).

We now turn our attention to this class of problems.

The *NP*-space

Failing to find close explicit upper and lower bounds on the complexity of a given problem, one may seek a proof that the complexity of the problem is related to that of some other problem; in other words, one attempts to classify the problem as being 'complete' in some larger class of problems. Such a result relates the complexity of the particular problem to that of the larger class as a whole.

One important class of problems, known as the *NP*-space problems, groups together the problems which are decidable, i.e. *solvable*, in principle, by computing means as we know them; *can be formulated* as a decision problem; *solvable in polynomial time* by a non-deterministic machine. The class *NP* includes an enormous number of practical problems that occur in business and industry (Garey and Johnson, 1979).

The term '*NP*-space problem' can be interpreted as 'non-deterministic polynomial time' problem, and, in turn, the class of problems solvable by a deterministic algorithm in polynomial time has been given the name of the *P*-space.

Relationship Between *P* and *NP* Spaces

The relationship between the classes *P* and *NP* is fundamental.

Every decision problem solvable by a polynomial time deterministic algorithm is also solvable by a polynomial time non-deterministic algorithm, that is $P \subseteq NP$. To see this, one simply needs to observe that any deterministic algorithm can be used as the checking stage of a non-deterministic algorithm. If $R \in P$, and AL is any polynomial time deterministic algorithm for R, we can obtain a polynomial time non-deterministic algorithm for R merely by using AL as the checking stage and ignoring the guess. Thus $R \in P$ implies $R \in NP$.

However, what we would like to know is, whether or not the subset $NP-P$ is empty, i.e. does $P=NP$ or not. This is considered to be the most important open problem in computer science today. There are currently two major lines of research into this dilemma.

(i) Proving that the problems of the NP-class are intractable, i.e. that the problems are so difficult that no polynomial complexity algorithm can possibly solve them. Unfortunately, at present, proving $P \neq NP$ is just as a hard as proving $P=NP$.

(ii) Examining the relationship between the NP problems. For example, within the class of NP a further subclass of problems is proven as NP-complete problems. A proof that a problem is NP-complete is usually considered a strong argument for abandoning further efforts to devise an efficient algorithm for its solution.

Complete problems have the property that all problems in the NP class (including NP-complete problems) are polynomially reducible to any one of them.

7.3 NP-complete Problems

In NP-space, problem $R1$ is called 'polynomially transformable' or 'polynomially reducible' to problem $R2$ if there is a polynomial time deterministic algorithm which transforms or 'reduces' $R1$ to $R2$. This is denoted '$R1 \propto R2$'.

Two problems $R1$ and $R2$ are called 'polynomially equivalent' if $R1 \propto R2$ and $R2 \propto R1$.

Problem R is called NP-complete if $R \in NP$ and every problem in NP is polynomially reducible to R. A proof that an NP-problem is NP-complete is a proof that the problem is not in P (does not have a deterministic polynomial time algorithm) unless every NP-problem is in P. In other words, all NP-complete problems are of equivalent difficulty: if one NP-complete problem can be always solved in polynomial time by a deterministic algorithm then all other problems in NP can be solved in polynomial time; if one problem in NP can be shown to require exponential time then all the others require exponential time.

The notion of the NP-complete problem was introduced in 1971 by Cook who proved the first-ever NP-complete problem, the so-called satisfiability problem.

If the satisfiability problem can be solved in polynomial time by a deterministic algorithm then $P = NP$.

In his proof, Cook followed the path essentially parallel to the path of Turing's earlier work on mathematical machines and their relation to problems of formal logic, and stated his proof in terms of the propositional calculus. In general, a sentence in the propositional calculus can be shown to be either true or false depending on which of its component statements are assumed to be true or false. Certain sentences, however, cannot be true under any interpretation because they are self-contradictory. Sentences that cannot be made true are said to be unsatisfiable.

Cook employed the propositional calculus to describe the operation of the non-deterministic Turing machines, used in the definition of the class *NP*. He showed that the calculations of any such machine can be described succinctly by sentences of the propositional calculus. When the machine is given a yes-instance of a problem in *NP*, its operation is described by a satisfiable sentence, whereas the operation of a machine given a no-instance is described by a sentence that cannot be satisfied. It follows from Cook's proof that if one could efficiently determine whether a sentence in the propositional calculus can be satisfied, one could also determine efficiently in advance whether the problem presented to a non-deterministic Turing machine will be answered yes or no. Since the problems in the class *NP* are by definition all those that can be solved by the non-deterministic Turing machine, one would then have an efficient method for solving all those problems. The sticky point, of course, is that there is no known efficient method of determining whether a sentence in the propositional calculus can be satisfied.

The satisfiability problem also known as 3-SATISFIABILITY is considered as one of the particularly celebrated problems in view of its historical role in founding the theory of *NP*-completeness, which undoubtedly is the most important development in computational complexity.

A year after Cook's paper Karp (1972) proved 21 problems were *NP*-complete, thus impressively demonstrating the importance of the subject. Independently of the West, the Russian school of computer scientists was working on a notion similar to that of *NP*-complete. In the Soviet literature the informal notion of 'search problem' was used in this context, and in 1973 Levin proved six problems *NP* complete, he called them 'universal search problems'.

Decision Problems

All problems considered in terms of transformation of one to another, are first formulated as a decision problem, i.e. in such a form that the solution to the problem is a YES/NO answer. These problems are then viewed as a recognition problem for a set of symbol strings, *B*.

For example, consider again the TSP.

Proving a Problem *NP*-complete

The principal technique used for transformation or reducing one problem to the other is a constructive transformation that maps any instance of the first problem into an equivalent instance of the second. Such transformation provides the means for converting any algorithm that solves the second problem into a corresponding algorithm for solving the first problem.

The general algorithm for proving problem R NP-complete may be outlined as follows:

(a) Formulate R as a decision problem.

(b) Choose an NP-complete problem, $R0$.

(c) Transform $R0$ to R.

(d) Show that the transformation function is of polynomial complexity.

It may be pointed out that, of course, the first proofs of the NP-complete problems were conceptually different (and much harder) as at the time there were yet no problems proved NP-complete.

A proof that the problem is NP-complete is normally interpreted as a proof that the problem is intractable. However, as we have seen there is a large number of problems which are termed 'apparently intractable' because for these problems no NP-completeness proof seem to be relevant.

Very recent developments in this direction concentrate on defining yet another class of problems, which would unify the apparently intractable problems as an identifiable research domain. For example, Valiant (1979a, b) defined the notion of $\#P$-completeness. Proving that a problem is $\#P$-complete shows that it is apparently intractable to compute, in the same way as proving a problem is NP-complete shows that it is apparently intractable to recognize: if a $\#P$-complete problem is computable in polynomial time, then $P = NP$.

The Complexity of Enumeration Problems

One problem which Valiant gave as an example of a $\#P$-complete problem is the computation of the permanent of a matrix. The permanent of a matrix is the variant of the determinant in which all summands are given positive signs. It is well known that the determinant is easily computable by Gaussian elimination, but many attempts over the past 100 years to find a feasible way to compute the permanent have all failed. Valiant proved that the problem is $\#P$-complete and thus for the first time convincingly demonstrated the reason for the failures.

The problem of the matrix permanent is an example of the large group of problems which are referred to as enumeration problems. An enumeration problem usually arises in connection with a search problem. In a search problem R, each instance has an associated solution set and we are normally required to find one element of the solution set, e.g. given an instance of the TSP we normally wish to find at least one optimum route while the instance may have several such routes. The corresponding decision problem asks whether or not the solution set is empty and will be fully satisfied with the answer YES or NO. The enumeration problem, on the other hand, poses the question: 'Given an instance of the search problem, what is the cardinality of the solution set, or, in other words, how many solutions are there to the instance?' Enumeration problems do not require the solutions, they merely determine how many there are, and are natural candidates for the type of problems that might be intractable even if $P = NP$.

Open Problems

By definition, a polynomial time non-deterministic algorithm checks in polynomial time a proposed string of symbols for a YES-solution. It may, however, not be able as 'quickly' to check for a NO-answer. The class *NP* is defined by the polynomial time non-deterministic algorithm for checking for a YES-solution only.

The problem for which a non-deterministic algorithm can be constructed such that it will check for a NO-solution in polynomial time, is said to belong to the class *CO–NP*. The class *CO–NP* consists of all problems that are the complement of some problem in *NP*. Intuitively, the problems in *NP* are of the form 'determine whether a solution exists', whereas the complementary problems in *CO–NP* are of the form 'show that there are no solutions'. It is not known whether *NP* = *CO–NP*, but there are problems that fall in the intersection *NP* ∩ *CO–NP*. An example of such a problem is the composite numbers problem: given an integer n, determine whether n is composite, i.e. there exist factors p and q such that $n = pq$, or prime. (Note that the problem of finding factors may be harder than showing their existence.)

The problems that belong to both *NP* and *CO–NP* classes are termed 'open'. Recently three important 'open' problems have been shown to be in the class *P*. The first is the linear programming problem shown to be in *P* by Khachian (1979). Prior to the publication of Khachian's paper all problems of linear programming had been solved using the Simplex method of Dantzig, or one of its variations, which are of exponential time complexity. Khachian has proposed an algorithm which solves the problem in polynomial time. The second is the problem of determining whether two graphs of degree at most d are isomorphic, shown to be in *P* by Luks (1980). Luks has shown that the algorithm is polynomial in the number of vertices for fixed d, but exponential in d. The third is the problem of factoring polynomials in one variable, shown to be in *P* by Lenstra (Lenstra *et al.*, 1982). The last result has later been generalized to polynomials in any fixed number of variables by Kaltofen (1982a, b).

7.4 Khachian's Algorithm

Linear programming is a subclass of the set of combinatorial optimization problems and as such it has a set of instances of the problem, a finite set of candidate solutions for each instance of the problem, and a function $f(x)$ that assigns, to each candidate solution of an instance, a solution value. The aim is to maximize/minimize the solution value for an instance of the problem.

The problem to be solved is to locate a vector x satisfying the system of inequalities:

$$a_{i1}x_1 + \ldots + a_{in}x_n \leq b_i, \qquad i = 1, \ldots, m \qquad (7.4.1)$$

for $n \geq 2$, $m \geq 2$, integer vectors **a** and **b** and real vector **x**.

Thus we are faced with locating an n-dimensional vector $\mathbf{x} \in R^n$.

Every now and then there appear innovations which, once thought of, appear simple and one wonders why one never thought of the idea before. Khachian's algorithm belongs to this class, its brilliance lying in its simplicity. To convey a comparison value between the Khachian algorithm and the Simplex method, the analogy was suggested to that of comparative standing of sequential and binary searches.

The Simplex method is somewhat like a sequential search in that it makes small local moves from one vertex to another, edge-connected, vertex and can conceivably, in the worst case, look at every vertex before finding the optimal one. Khachian's algorithm relates to the binary search method in that it divides up the search space until the answer is homed in on. The algorithm centres around defining an initial finite search space within which the solution $\mathbf{x}$, if it exists, is known to lie. This search space is then divided in two and by the use of the dividing criterion, the half in which the vector $\mathbf{x}$ does not lie is discarded. Next, a new smaller search space, containing the previous half-space, is computed and the process is repeated until $\mathbf{x}$ is located. Thus one is faced with the problems of:
(a) how to define the initial search space;
(b) what criterion to use to divide the space;
(c) how to define the resulting new search space;
(d) how to recognize a solution $\mathbf{x}$.
Khachian's answer to these problems was to use a series of monotonic decreasing ellipsoids.

At the time, when Khachian published his algorithm which solves linear programming deterministically in polynomial time, it was believed that the problem was a member of the *NP*-complete group. So the result was greeted with great excitement and sensation as one believed that the dilemma of $P = NP$ has finally been cracked and all other practically important difficult problems such as code-cracking and the TSP were now within the scope of our present technology. However, a closer study of the problem's properties revealed that linear programming has the properties of both groups, the *NP* and *CO–NP*.

The theory, however, assumes that an *NP*-class problem cannot be a member of the *CO–NP*-class as well because there is strong evidence that *NP* $\neq$ *CO–NP*. The evidence is, of course, not a rigid proof, but nevertheless it is strong enough for linear programming to be excluded from the *NP*-class. More detailed logic behind this exclusion is given in Garey and Johnson (1979).

7.5 Beyond the *NP*-class

There are many important problems which do not seem to be solvable in polynomial time even by a non-deterministic machine. For example, the problem of proving that a Boolean expression is always 1 (or 0) does not seem to be possible to solve in polynomial time with a non-deterministic algorithm. Another problem in this category is the so-called *placement problem*, i.e. placing a set of circuits such that the total length of wires is less than some specified value L, and proving that there exists no placement with a wire length less than L; this problem does not seem to be possible to solve in a non-deterministically polynomial time.

In order to give some sort of classification to such problems and in this way to unify them as a research domain, the concept of an 'oracle machine' has been developed. One such notion of an 'oracle machine' is the query machine of Cook (1971).

A query machine is a (non-deterministic or deterministic) Turing machine with a distinguished work tape called the query tape and three distinguished states, the query state, the yes state and the no state. The computations of a query machine depend not only on the input but also on a given set of words called the oracle. The actions of a query machine with oracle B are identical to those of Turing machines with one exception. If the machine enters its query state at some step, the machine next enters its yes state if the non-blank portion of the query tape contains a word in B; otherwise the machine next enters its no state. An oracle machine M operates within time $T(n)$ if and only if for every input x, every computation of M (relative to any oracle) halts within $T(|x|)$ steps.

This allows one to develop a hierarchy of machines, each of which is 'complete' in a higher level of language. Meyer and Stockmeyer (1972) observed that this process of defining new complexity classes of problems in terms of old ones could be extended indefinitely, yielding classes of greater and greater apparent difficulty. We thus obtain the polynomial hierarchy or the *P*-hierarchy. The polynomial hierarchy is defined in terms of polynomial bounded oracle machines (Stockmeyer, 1979).

Polynomial Space Completeness

In our analysis of the computational complexity classes so far the emphasis has been on just one of the major 'resources' required by a computation, the time it takes the computation to be performed. The amount of computer memory required by the computation is often just as important. The class of all problems solvable in polynomial space is termed the *PSpace*.

All problems solvable non-deterministically in polynomial time can be solved in polynomial space, so the *PSpace* includes *NP* and *CO–NP* (Donath, 1979).

However, the question whether there exist problems solvable in polynomial space that cannot be solved in polynomial time remains unresolved. It is conjectured that *PSpace* may contain the problems that are thought by some to be harder than the problems in *NP* and *CO–NP*.

An even more powerful conjecture suggests that there exists a different set of problems complete in *PSpace* (Aho *et al.*, 1974). The *PSpace*-complete

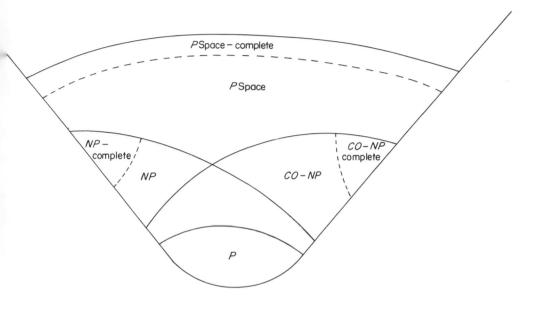

Figure 7.5.1   Complexity classes

problems are the problems in *PSpace* such that if any one of them is in *NP*, then *PSpace* = *NP*, or if any one is in *P*, then *PSpace* = *P*. In Fig. 7.5.1 are shown some important complexity classes and their possible relationships.

Another purpose of the polynomial hierarchy is to design a more detailed way of classifying the so-called *NP*-hard problems, the name given to the problems which are known to be at least as 'hard' as an *NP*-complete problem, to which an *NP*-complete problem can be transformed, but which themselves may or may not be members of the *NP*-class. The notion of the *NP*-hard problem is general enough to allow for the problems other than just decision problems to be proved to be at least as hard as the *NP*-complete problems. When a new problem is encountered, which is an apparently hard problem, one first asks the question of whether the problem is in the hierarchy at all; some ways of determining this have been developed. For further reading on this subject the reader is referred to the work by Karp (1972) and Stockmeyer (1977).

8

The Theory of Computational Complexity

In the world of computing the critical question about a problem P is not 'Is P solvable?', but rather 'Is P computable in practical terms, that is, is there an algorithm for P which will solve it in the time and space we have available?'

Intuitively one feels that besides the skill of designing a 'good' algorithm and the strength of contemporary computers there is an additional factor to be considered, which has to do with the individual properties of the problem itself. Computational complexity has been developed in order to study such questions and to aid these more practical aspects of solving problems.

As we have seen on a variety of complexity measures, each measure is specifically suited for the particular class of problems. Whenever possible, such a measure is designed so as to project the most essential features of the problem and subsequently to give quantitative estimates of these basic features, e.g. the matrix problem is about multiplication of numbers and the complexity measure counts the number of these multiplications needed to obtain the solution; the sorting problem within the specific definition is about comparison of the elements of a given set and the measure counts the number of these comparisons required to determine the result; the graph problem is concerned with both the number of vertices in the given problem and the number of edges, and so the complexity measure is designed to 'measure' both parameters, etc.

In parallel with the measures which are designed to assess the complexity of a specific problem, a general theory of computational complexity has evolved with deeper insight into the question of the intrinsic difficulty of computing a problem. The approach is motivated by specific complexity measures and their relations, but the general theory is not about real computers and practical algorithms run on such computers; instead it studies computational complexity measures which are defined for all possible computations and which assign a complexity to each computation that terminates.

In order to achieve this level of abstraction of the complexity notion, one needs a way effectively to specify all possible algorithms. Then the complexity measure will show how many 'steps' it takes to evaluate any one of these algorithms on any specific argument. We will give a brief account of some significant results in this direction.

8.1   Some Basic Definitions

The following definitions which are used in the complexity theory are helpful
in the subsequent summarized exposition of the theory.

A *computable function* (problem) is a function whose values can be
calculated in some kind of effective way.

A *recursive function* (problem) is a function which is defined by specifying
each of its values in terms of previously defined values, and possibly using
other previously defined functions.

A *primitive function* (problem) is a recursive function in the definition of
which its previously defined values are used once only, e.g.

$$F(0) \quad = 1, \quad F(1) = 1,$$
$$F(n+1) = F(n-1) + F(n), \tag{8.1.1}$$

i.e. the problem of computing the Fibonacci numbers is a primitive problem
(function).

A *second-order* (or higher-order) *recursive function* (problem) is a recursive function in the definition of which its previously defined values are used
recursively, e.g.

$$\Psi(0, y) \qquad = y+1,$$
$$\Psi(x+1, 0) \quad = \Psi(x, 1), \tag{8.1.2}$$
$$\Psi(x+1, y+1)= \Psi(x, \Psi(x+1, y)),$$

i.e. the problem of computing the Ackerman function is the second-order
problem (function).

A *partial function* (problem) is a function defined over a finite domain of its
argument.

A *total function* (problem) is a function defined over the whole infinite
domain of its argument.

A *partial recursive function* (problem) is a recursive function defined over a
finite domain of its argument. We shall use the terms 'recursive function' and
'total computable function' interchangeably (for the proof of this fact see, for
example, Cutland, 1980).

The set of natural numbers as usual is denoted by $N = \{0, 1, \ldots\}$.

Definition   Let A be a subset of N. The *characteristic function* of A is the
function c_A given by

$$c_A = \begin{cases} 1, & \text{if } x \in A, \\ 0, & \text{if } x \notin A. \end{cases} \tag{8.1.3}$$

The A is said to be recursive if c_A is computable. Recursive sets
are also called computable sets.

Definition   Let A be a subset of N. Then A is *recursively enumerable* if the
function f given by

$$f(x) = \begin{cases} 1, & \text{if } x \in A, \\ \text{undefined}, & \text{if } x \notin A, \end{cases} \qquad (8.1.4)$$

is computable. The phrase 'recursively enumerable' is almost universally abbreviated r.e.

8.2 Theoretical Complexity Measures and Related Results

Consider (a recursive enumeration of) all Pascal programs. The *complexity of the ith (Pascal) program on argument n* is defined as the number of instructions (call an instruction a 'step') executed before the program halts on input n. An important fact to be noted in this example is that the measure is associated with the (Pascal) programs or algorithms and not directly with the functions they compute. The reason for this is that in computations one normally deals with algorithms which specify functions (in order to compute them), and since for each computable function there are infinitely many algorithms which compute it, a measure counting the number of computational steps required to compute the function would not be meaningful if associated with the function itself. Furthermore, one cannot either define the complexity of a function as that of its 'best' algorithm because there exist functions which have no 'best' algorithm (see the speed-up theorem later in the text; also Hartmanis and Hopcroft, 1971).

Based on the Turing machine approach, one can define measures such as 'the number of steps needed to perform a Turing computation' and 'the amount of tape used to perform a computation'.

In general a computational complexity measure is defined as a (recursive) set of algorithms which compute all (partial recursive) functions, where to each algorithm a step-counting function is assigned, that gives the amount of resource used by a given algorithm on a specific argument.

So, let AL_1, AL_2, ... be the set of algorithms and S_1, S_2, ... the set of corresponding step-counting functions, then within the computational complexity measure the following two conditions hold:
1. The algorithm $AL_i(n)$ is defined if and only if $S_i(n)$ is defined.
2. For any given number of steps m and any algorithm AL_i working on argument n, one can determine (recursively) whether $AL_i(n)$ halts in m steps, i.e. whether $S_i(n) = m$. In other words, if the ith Pascal program halts on input n, then the number of instructions (steps) executed before halting is well defined. However, if the ith Pascal program does not halt on input n, then we cannot determine how complex the computation is since the measure is not defined. What one can do for each i and n is to determine whether the ith Pascal program halted on input n after execution of m instructions (steps) for any given n. This is achieved in an obvious way by just performing m instructions of the ith program on input n and noticing whether the computation halts on the last instruction. The following is a formal definition of a computational complexity measure.

Definition   A computational complexity measure Φ is an admissible enumeration of the partial recursive functions (algorithms) AL_1, AL_2, ... to which are associated the partial recursive step-counting functions S_1, S_2, ... such that

(i) $AL_i(n)$ is defined if and only if $S_i(n)$ is defined;

(ii) $P(i,n,m) = \begin{cases} 0, & \text{if } S_i(n) \neq m, \\ 1, & \text{if } S_i(n) = m \end{cases}$ is a recursive function.

The definition is general enough and many natural measures satisfy this definition, particularly if we note that given a set of step-counting functions, any recursive function $f(n)$, $f(n) \geq n$, can be applied to each step-counting function to obtain a new set of step-counting functions.

At the same time, the definition of a computational complexity measure is restrictive enough to eliminate as step-counting functions those functions which in no real sense measure the complexity of the computation. For example, the number of recursions used to define a function in a schema for primitive recursion cannot be used for step-counting functions since the schema is not capable of representing all partial recursive functions, and thus one does not have an admissable enumeration of all algorithms.

Below we give a brief summary of the results implied by the definition of computational complexity; these are presented mostly without proof (proofs may be found in Hartmanis and Hopcroft, 1971).

(a) *For any computational complexity measure there exist arbitrary complex (total) computable functions.*

Theorem 8.2.1 Let Φ be a computational complexity measure and f any computable function. Then there exists a (total) computable function g such that, for any index i for g, $S_i(n) > f(n)$ for infinitely many n.

(b) *For any computable f there exist (total) computable functions whose complexity exceeds f almost everywhere. There exist arbitrary complex bounded functions.*

Theorem 8.2.2 Let Φ be a complexity measure. Then for any computable function f there exists a (total) computable function g such that for any index i for g, $S_i(n) > f(n)$ for almost all n.

The theorem is followed by a corollary:

There exist arbitrary complex $0-1$ valued functions in all measures.

(c) *There can be no recursive relation between functions and their complexities, e.g. a bounded function does not imply a bound on its complexity.*

Theorem 8.2.3 Let Φ be a complexity measure and g a (total) computable function. Then there does not exist a recursive function k such that for each algorithm AL_i that computes g on input n,

$$k(n, g(n)) \geq S_i(n) \text{ almost everywhere (a.e.)},$$

and there does exist a recursive function h such that for each algorithm AL_i that computes g on input n,

$$h(n, S_i(n)) \geq g(n) \quad \text{a.e.}$$

The second part of the theorem asserts that one can bound recursively the size of any computable function by its complexity. This leads to the conclusion that the superfast growing computable functions are 'supercomplex'.

(d) *Although there is no recursive relation between the value of a function and its complexity, there is a recursive relation between the complexity of an algorithm in any two measures; a function which is 'easy' to compute in one measure is easy to compute in all measures, and furthermore, the complexity of any class of functions which is recursively bounded in one measure, e.g. polynomials, primitive functions, etc. is recursively bounded in every complexity measure.*

Theorem 8.2.4 Let Φ and $\hat{\Phi}$ be complexity measures and g a (total) computable function. There exists a recursive r such that, for any two algorithms AL_i and $\hat{A}L_i$ that compute g on input n,

$$r(n, S_i(n)) \geq \hat{S}_i(n) \quad \text{and} \quad r(n, \hat{S}_i(n)) \geq S_i(n) \quad \text{a.e.}$$

8.3   The Speed-up Theorem

We have noted earlier that it is not possible to define a complexity of a function based on its 'best' algorithm since there exist functions which have no 'best' algorithm. This curious result is derived from the Speed-up theorem of Blum (1967). The theorem shows that in any complexity measure there exist functions whose computation can be arbitrarily sped up by choosing more and more efficient algorithms.

Definition   A predicate $M(n)$ holds for almost all n, or almost everywhere, often written as a.e., if $M(n)$ holds for all but finitely many natural numbers n, in other words, if there is a number n_0 such that $M(n)$ holds whenever $n \geq n_0$.

Consider the following example. AL_1 and AL_2 are two algorithms for computing a total function (problem) g such that their complexities S_1 and S_2 for any x satisfy the following:

$$3(S_2(x)) < S_1(x).$$

We naturally say that algorithm AL_2 performs three times as fast as algorithm AL_1. In terms of the Speed-up theorem, for a particular instance, one would say that there is a total function g such that if AL_1 is any algorithm for g, then there is another algorithm for g that is more than three times as fast on almost all inputs. Thus, in particular, there is no best algorithm for computing g.

The example represents speed-up by a factor of 3, given by a computable function $r(x) = 3x$. The Speed-up theorem will give speed-up by any preassigned computable function. An amazing implication of the Speed-up theorem is the observation that no matter which two universal computers (a universal computer is the computer which, over a given alphabet, embodies all other computers that can be defined over the same alphabet) we select, no matter how much faster and more powerful one of the machines is, functions exist which cannot be computed any faster on the more powerful machine. This is because for any algorithm which one uses to compute such a function on the more powerful machine, according to the theorem another algorithm exists which is so fast that even on the slow machine it runs faster than the other algorithm on the faster machine.

The first proof of the Speed-up theorem by Blum uses some complex results from the computability theory. Later Hartmanis and Hopcroft (1971) gave a more direct proof of the theorem using the technique of diagonalization. We shall outline the proof on the lines of Hartmanis and Hopcroft.

The Speed-up theorem Let Φ be a complexity measure and $r(n)$ a recursive function. There exists a total computable function $g(n)$ so that for each algorithm AL_i that computes g there exists another algorithm AL_j which also computes g, such that $S_i(n) \geq r(S_j(n))$ a.e.

Comments preceding the proof. Hartmanis and Hopcroft in their proof of the theorem first observe that the Speed-up theorem is measure-independent, i.e. if it holds in any measure it holds in all measures. Then they prove the theorem directly for the well-understood complexity measure of tape-bounded Turing machine computations. We shall outline only a simpler variant of their proof. However, first some useful related information and the detailed set-up for the proof are in order.

Let Φ, a complexity measure, consist of all possible algorithms $\{AL_i\}$ for solving a certain problem, e.g. a set of (partial recursive) algorithms, $\{AL_i\}$, is considered for the sorting problem, which maps a set of n integers into an ordered set of the same n integers, and a corresponding set of step-counting functions, $\{S_i\}$.

Further, let the algorithm AL_i be a one-tape Turing machine TM_i and the step-counting function be the number of tape cells used by TM_i on input n, denoted by $L_i(n)$. The definition of a one-tape Turing machine is extended so as to make the model adequate for the computational purpose in hand, e.g. that the machine is assumed to never cycle on a finite segment of its tape, that the $L_i(n)$ of the ith Turing machine is larger than the length (the number of tape cells) used for the machine's description, and that the ith Turing machine has at most i different tape symbols. Hence the proposed complexity measure consists of a set of one-tape Turing machines $\{TM_i\}$ and a set of the corresponding counting functions $\{L_i\}$.

Next, let a tape-constructable computable function $f(n)$ be defined as a function which exists if and only if there exists a one-tape Turing machine which uses exactly $f(n)$ tape cells for input n, $n = 1, 2, 3, \ldots$.

The proof involves designing, by a diagonal process, a function ϕ which cannot be computed quickly by 'small' machines. The diagonal procedure is formulated as follows.

Let $\{\phi_i\}$ be a set of total computable functions so that function $\phi_i(n)$ is computed by algorithm AL_i on input n. Then for a properly chosen function h, if the ith machine computes $\phi_i(n)$ on fewer than $h(n-i)$ tape cells then ϕ is different from ϕ_i.

Lemma Let $r(n)$ be any recursive function. There exists a total computable function $\phi(n)$ such that for each algorithm AL_i which computes $\phi(n)$ there exists another algorithm AL_j which also computes $\phi(n)$ such that for the step-counting functions L_i and L_j,

$$L_i(n) \geq r(L_j(n))$$

for sufficiently large n.

Proof. Define a strictly increasing tape-constructable function $h(n)$ as follows:

$$h(n) = \begin{cases} 1, & n = 1, \\ r(h(n-1))+1, & n > 1. \end{cases} \qquad (8.3.1)$$

Note that by definition

$$h(n) > r(h(n-1)) \qquad (8.3.2)$$

and h is tape-constructable because $r(x)$ and $r(x)+1$ are tape-constructable.
Next, define the function $\phi(n)$ so that:
(i) $\phi_i(n) = \phi(n)$ implies that

$$L_i(n) \geq h(n-i) \text{ a.e.} \qquad (8.3.3)$$

(ii) for each k there exists an index j such that

$$\phi_j(n) = \phi(n) \quad \text{and} \quad L_j(n) \leq h(n-k). \qquad (8.3.4)$$

So, defined function ϕ ensures the result of the lemma, that is, that for any algorithm TM_i for ϕ, there exists another algorithm TM_j for ϕ with

$$L_i(n) > r(L_j(n)) \text{ a.e.} \qquad (8.3.5)$$

This is achieved by choosing TM_j so that

$$L_j(n) \leq h(n-i-1), \qquad (8.3.6)$$

which, together with equations (8.3.3) and (8.3.2), gives

$$L_j(n) \geq h(n-i) > r(h(n-i-1)) > r(L_j(n)). \qquad (8.3.7)$$

Using definition (8.3.1), the function is constructed as follows. For $n = 1$,

$$\phi(1) = \begin{cases} \phi_1(1) + 1 & \text{if } L_j(1) < h(1), \\ 0, & \text{otherwise.} \end{cases} \qquad (8.3.8)$$

If $L_i(1) < h(1)$, cancel the first Turing machine from the list of Turing machines.

For $n = 2, 3, \ldots$,

$$\phi(n) = \begin{cases} \phi_i(n) + 1, & \text{where } i \leqslant n \text{ is the smallest index not} \\ & \text{already cancelled such that} \\ & L_i(n) < h(n-i), \text{ and cancel index i;} \\ 0, & \text{if no such } i \text{ exists.} \end{cases} \quad (8.3.9)$$

For $\phi(n)$ so constructed, if the ith Turing machine computes $\phi(n)$, that is, $\phi(n) = \phi_i(n)$, then

$$L_i(n) \geqslant h(n - i) \quad \text{a.e.} \quad (8.3.10)$$

since for sufficiently large n_0, each j which will eventually be cancelled has already been cancelled, and thus if

$$L_i(n) < h(n - i) \quad (8.3.11)$$

for any $n > n_0$, then

$$\phi(n) = \phi_i(n) + 1 \neq \phi_i(n), \quad (8.3.12)$$

that is, $\phi(n)$ cannot be computed by TM_i. As far as the second part of the definition of $\phi(n)$ is concerned, for each k there exists TM_j which computes ϕ in

$$L_j(n) \leqslant h(n-k) \quad (8.3.13)$$

tape cells.

Here, TM_j operates as follows. Think of a value v, such that:
 (i) for each i, $i \leqslant k$, whichever gets cancelled, gets cancelled for a value of $n < v$;
(ii) for each $n \leqslant v$, TM_j has stored in its finite control the value of $\phi(n)$ and simply prints out the appropriate value;
(iii) for $n > v$, TM_j computes the smallest i not already cancelled by
    :- laying off $h(n-k)$ tape cells;
    :- storing in finite control, the values of i cancelled on input n, $n \leqslant v$;
    :- for $v < m \leqslant n$, simulating each TM_i, $k < i \leqslant n$, determining which machine gets cancelled at each value of $m < n$, then finding the smallest uncancelled i such that $L_i(n) < h(n-i)$ and setting $\phi(n) = \phi_i(n) + 1$.

If no i exists, $\phi(n)$ is set equal to zero. The simulation requires $h(n-k)$ tape cells since TM_j simulates only machines with indices greater than k and simulates the machine i only until it exceeds $h(n-i)$ tape cells. Since machine TM_i has at most i tape symbols, the simulation requires at most

$$ih(n-i) < h(n-k)$$

tape cells.   QED.

Proof of the Speed-up theorem. Without loss of generality, let us assume that $r(n)$ is an increasing monotonic function. Since all measures are recursively related, there exists a strictly monotonic unbounded R such that

$$L_i \leqslant R(S_i) \quad \text{and} \quad S_j \leqslant R(L_j) \quad \text{a.e.} \tag{8.3.14}$$

provided S_i and L_j grow faster than n. Now set

$$\hat{r}(n) = R(r(R(n))). \tag{8.3.15}$$

Select ϕ by lemma so that for each i such that $\phi_i = \phi$ there exists TM_j for ϕ with

$$L_i \geqslant r(L_j) \quad \text{a.e.} \tag{8.3.16}$$

Then

$$R(r(S_j)) \leqslant R(r(R(L_j))) \leqslant L_i \leqslant R(S_i) \text{ a.e.} \tag{8.3.17}$$

But $R(r(S_j)) \leqslant R(S_i)$ a.e. and R strictly monotonic implies

$$r(S_j) \leqslant S_i \quad \text{a.e.} \tag{8.3.18}$$

as was to be shown.   QED.

It remains to be noted that not all computable functions can be sped up. One should also observe that the speed-up is not effective in that the value of v in the construction of the lemma is not effectively determined. It can be shown that the Speed-up theorem is not in general effective. Other proofs of the theorem can be found in Cutland (1980) and Young (1973).

8.4   Complexity Classes. The Gap Theorem

Earlier we postulated two properties which any complexity measure must have: it must have a set of computable functions (algorithms) and a set of the corresponding step-counting functions, i.e. $\Phi = \{\phi_i(n)\}$ and $\{S_i(n)\}$.

We now consider the classes of functions whose complexity is bounded by recursive functions.

Suppose that f is any recursive function. From the point of view of complexity, a natural class of functions comprises those functions having step-counting functions bounded by f. Hence the complexity class can be defined as

$$C_f^\Phi = \{\phi_j(n)/S_j(n) \leqslant f(n) \text{ a.e.}\} \tag{8.4.1}$$

The class C_f^Φ or C_f as thus defined is the complexity class of f relative to the set of step-counting functions $\{S_j(n)\}$.

Taking another recursive function $g(n)$ we get a complexity class C_g^Φ. Suppose the function $g(n)$ is such that $g(n) > f(n)$ for all n, e.g. $g(n) = 2f(n)$.

Since the set of step-counting functions in C_g^Φ is bounded by g (and not by f), one would intuitively expect the class C_g^Φ to contain some new functions which are not in C_f^Φ, especially if g is much larger than f. One can, for example, construct a new complexity class, C_g^Φ, by adding to the C_f^Φ a new function $\phi(n)$ such that its step-counting function is bounded by g (but not by f). Such a result would mean that $C_g \supset C_f$.

However, this intuition has been proven wrong. In fact it is shown that in *any complexity measure one can find f and g with g greater than f by any preassigned computable factor, such that $C_g^\Phi = C_f^\Phi$.* The f and g can be chosen so that there is no step-counting function that lies between f and g for more than finitely many n. This remarkable theorem is called the Gap theorem and is due to Borodin (1969). It asserts that *for any complexity measure the step-counting functions are sparse relative to recursive functions.*

Comments

Theoretical computer science is concerned with the major problems of understanding the nature of computation and its relationship to various existing computing methodologies. This requires development of powerful mechanisms for abstraction and deduction, particularly in the light of new advances in computing methodologies involving concurrent processes. Development of new concepts and ideas is urged by the new technological progress in computing. In the chapters that follow we present and analyse some of the new computational methods.

Part Two
Parallel Algorithms

9

Computational Complexity of Parallel Algorithms

9.1 Introduction

In the last three decades of computing, we have witnessed truly impressive advances in the speed, size, and sophistication of computers. Early computers were designed to perform programming instructions sequentially, the same way as human beings do. However, different from humans, computers execute these instructions much much faster. And still even the fastest sequential computers are constrained by circuit speeds limits; 10^9 operations per second is still beyond the reality on a serial computer of today. But even allowing for the most advanced computer construction we cannot expect to build computers that are significantly faster than those we already have, since electrical signals take about 1 nanosecond (10^{-9} sec) to travel 1 foot and even the smallest computers will have some components separated by at least an inch or two, hence somewhere around 10^{10} operations a second must be the limit on physical grounds (Churchhouse, 1983).

The next step in searching for methods for improvement in computation speed is to allow parallelism or concurrency. A parallel or concurrent algorithm is an algorithm which permits carrying out many operations simultaneously or in parallel. Basically it means the utilization of the fact that very large computational and information-processing problems can be partitioned in such a way that various parts of the work can be carried out independently and in parallel, and the results combined together when all subcomputations are complete.

The concept of parallelism can be pursued in a hardware-orientated direction when a parallel architecture of a specific kind is built and the parallel algorithms for solving different problems are developed to make use of these parallel hardware features to the best advantage. Other approach can be called a problem-orientated computation parallelism when the fundamental question of whether and, if so, how much the parallel algorithms can truly enhance the speed of obtaining a solution to a given problem.

The earliest parallel computer was built in 1969 (Illiac, USA) and today, in the mid 1980s, several parallel computers are commercially available; the most powerful parallel computer, Cyber 205, manages 10^9 floating-point operations per second (Purcell, 1982). Parallel processing has firmly moved into the lives of programming professionals, practising scientists, engineers, and alike. It has become increasingly clear that the most powerful computer of the future will not be sequential but parallel.

The usefulness of a parallel computer depends greatly on the invention of suitable parallel algorithms that operate efficiently on such a computer, and on the design of parallel languages in which these algorithms can be expressed. Hence a major rethinking in these areas is now under way.

The parallel computers which exist in practice have various restricting factors dictated by practical limitations; due to these limitations the performance of parallel algorithms is often slowed down, compared with the algorithm's 'ideal' performance on an 'ideal' parallel computer. However, theoretically, abstracting from all practical limitations, we can formulate a basic question that refers to the inherent parallelism in the problem: which problems can be solved substantially faster using many processors rather than one processor? This fascinating question has been formalized by Nicholas Pippenger (1979) who defined the class of problems solvable 'ultra-fast on a parallel computer with a feasible amount of hardware'. This class of problems is referred to as NC for 'Nick's class'. The class NC is defined so that it remains independent of the particular parallel computer model chosen. The role of the NC is similar to that of the NP-complete set of problems; it allows the classification of a great many problems in terms of their parallelism.

In this chapter we shall define the major concepts of parallelism, introduce some parallel processing algorithms for numerical and non-numerical computational problems, and investigate the current views on inherent parallelism in a problem. Research into the inherent parallelism of a problem is considered the most timely activity of today since at any moment of time the ever advancing technology would become capable of constructing and building the parallel systems 'to order', and so to allow the parallel algorithms developed to be used in the most efficient way.

9.2. Parallel Computers

At the earliest stages of the development in parallel computer concepts, four categories of computer systems were distinguished (Flinn, 1966). The three computer architectures are illustrated in Fig. 9.2.1.

The earliest sequential computers, of the SISD type, which executed a single instruction at a time, using a single piece of data, were well suited to the technology of the times; their architecture was straightforward. With the progress of technology, computer users required greater performance and consequently the SISD machines were made faster and faster, using newer and better components and designs. However, the inherent limitation of the SISD architecture—the performance of only one instruction at a time—remained.

A fresh approach to increasing the speed of computation was to make multiple copies of portions of the SISD hardware. In the SIMD architecture all major portions of the hardware—the operand fetch, execution, result store—are replicated, so that the execution of a single instruction enables several values to be fetched, computed upon, and answers stored. For a great many problems, this provides a substantial performance improvement. With sufficient hardware, entire vectors of numbers can be operated upon simultaneously.

The two major parallel architectures of today, of the SIMD type, are an array processor and a vector processor. The two architectures are outlined in the sequel.

Array Processor

An array processor is an aggregate of identical processors each constrained to execute the same instruction simultaneously on local data. A block diagram of an array processor system is illustrated in Fig. 9.2.2(a).

In the system shown in Fig. 9.2.2, the control processor is itself a computer. It has its own local memory, arithmetic unit, registers, and control unit. The control unit fetches an instruction and determines whether or not it is a multiple-data operation. If it is not a multiple-data operation, the control processor executes the operation. If it is a multiple-data operation, the instruction is passed on to the arithmetic processors, with each processor holding the part of the multiple data being operated on inside its local memory.

In the simplest case, the arithmetic processors hold exactly one element of the mutliple-data set, but if the size of the multiple-data set exceeds the number of processors in the array then the processors may hold a subset of the multiple-data set.

All arithmetic processors receive instructions from the control processor simultaneously, the entire multiple-data operation proceeding in parallel. The arithmetic processors operate fully under control of the control processor; in effect, they are 'slaves' of the control processor. The crucial difference between the control processor and the arithmetic processors is that the arithmetic processors lack the ability to interpret conditional branch instructions.

With regard to the memory arrangement within such a multiprocessor system, the two major architectures are distinguished—a loosely coupled and a tightly coupled systems. The two systems are illustrated in Fig. 9.2.2(b).

In a tightly coupled multiprocessor system the memory is common to all processors, who share it when and as needed; for practical convenience the shared memory may be partitioned into independent and interleaved memory modules with full connection existing between the processors and the memory modules. In a loosely coupled multiprocessor each processor has its own memory and the communications between the processors are achieved through the interconnection network. The processors which have their own local memories are called the processing elements or simply PEs. In some

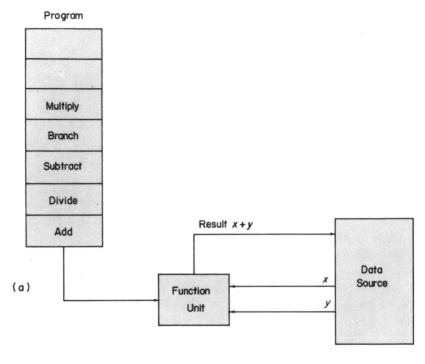

(a)

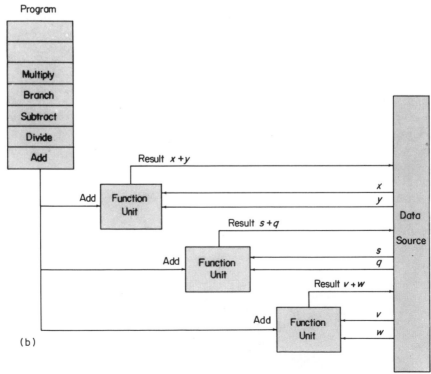

(b)

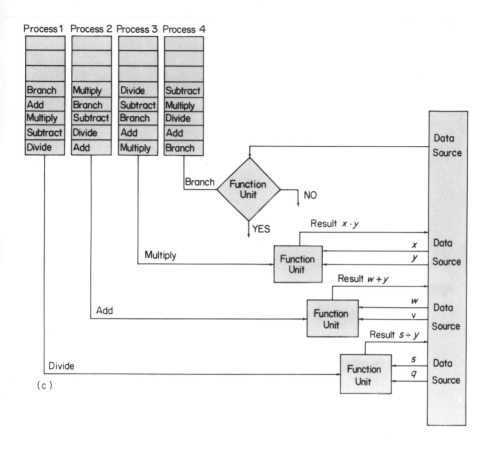

Figure 9.2.1 (a) SISD model = single instruction single data stream. A single instruction is executed at a time, using a single piece of data. (b) SIMD = single instruction multiple data stream. The SISD portions of the hardware are replicated so that the execution of a single instruction causes several values to be fetched, computed upon, and the answers stored. (c) MIMD = multiple instruction multiple data stream. The MIMD architecture can yield high performance at low hardware cost by keeping all processor hardware utilized executing multiple parallel programs simultaneously. For example, while a division is in progress for one process, an addition may be executing for another, a multiplication for a third. Because the multiple instructions executed concurrently by a MIMD machine are independent of each other, execution of one instruction does not influence the execution of other instructions and processing may be fully parallel at all times

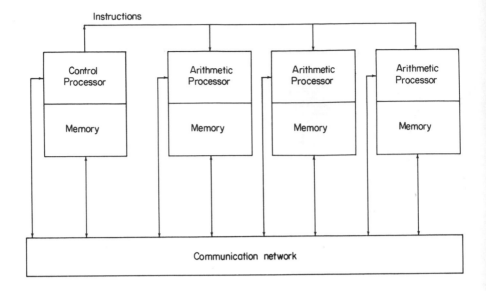

Instructions

| Control Processor | Arithmetic Processor | Arithmetic Processor | Arithmetic Processor |
| Memory | Memory | Memory | Memory |

Communication network

(a)

Figure 9.2.2   An array processor diagram. (a) A block diagram of an array processor system. (b) Two major multiprocessor systems: (i) tightly coupled multiprocessor— the interconnection network (IN) connects the processors to common memory modules; (ii) loosely coupled multiprocessor—a processing element (PE) consists of a processor and a memory module

loosely coupled systems the PEs are intelligent enough to execute their own programs. In other systems, e.g. the ICL DAP (Parkynson, 1982), the arithmetic processors are organized as an array of identical single-bit PEs.

The PEs within the system can be linked together in a variety of ways, with the emphasis on the PEs relative independence or dependence.

Processor independency

Each processor receives a subset of data into its local memory. The control processor then simply issues a single instruction or a sequence of instructions to be performed by all processors simultaneously but on different data stored in their local memories. Suppose that the matrix addition of two square matrices of size n, $\mathbf{B}$, and $\mathbf{C}$ is performed by n processors and the result stored in matrix $\mathbf{A}$. Each processor holds a row of the operand matrices as illustrated in Fig. 9.2.3. The control processor issues a loop of instructions of the form

$$\mathbf{A}(k) = \mathbf{B}(k) + \mathbf{C}(k), \qquad k = 1, \ldots, n.$$

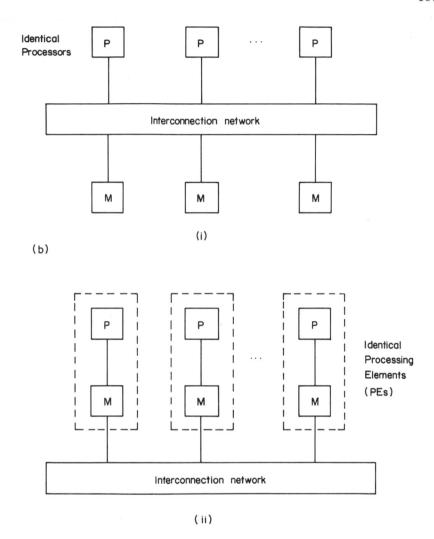

Identical Processors

Interconnection network

M M M

(i)

(b)

Identical Processing Elements (PEs)

P P P

M M M

Interconnection network

(ii)

(b)

If $n \times n$ processors were available for the computation then each processor would hold just a particular subscript, (i, j), of the operands $\mathbf{B}$ and $\mathbf{C}$:

$$\mathbf{A}(i, j) = \mathbf{B}(i, j) + \mathbf{C}(i, j), \qquad \begin{aligned} i &= 1, \ldots, n, \\ j &= 1, \ldots, n, \end{aligned}$$

and the control processor would simply issue the command

$$\mathbf{A} = \mathbf{B} + \mathbf{C}$$

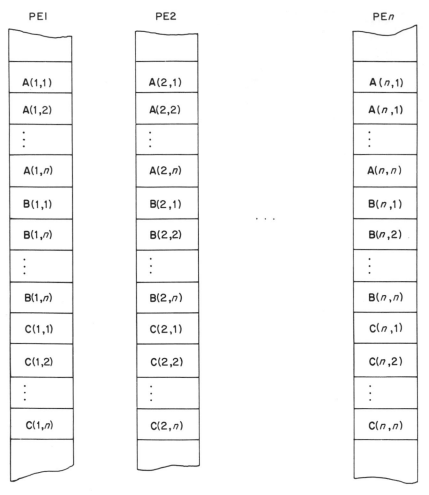

Figure 9.2.3 The memory arrangements for n processors PE1 to PEn, each holding a
row from matrices A, B, and C

to accomplish the matrix addition. That is, in this set-up the processors do not
communicate with each other, only with the control processor. In general,
since the large number of processors are working in parallel, such a scheme is
considered to be very efficient.

Processor dependency

Suppose that we want to find the scalar product of two vectors, $\mathbf{x}$ and $\mathbf{y}$,

$$z = \sum_{i=1}^{n} x_i y_i \, .$$

With n processors, all n products, $x_i y_i$, can be formed simultaneously in a way, analogous to the previous mode. Then the summation can be performed by adding operands as pairs, e.g.

$$x_1 y_1 \quad x_2 y_2 \quad x_3 y_3 \quad x_4 y_4 \quad x_5 y_5 \quad x_6 y_6 \quad x_7 y_7 \quad x_8 y_8$$
$$x_1 y_1 + x_2 y_2 \quad x_3 y_3 + x_4 y_4 \quad x_5 y_5 + x_6 y_6 \quad x_7 y_7 + x_8 y_8$$
$$(x_1 y_1 + x_2 y_2) + (x_3 y_3 + x_4 y_4) \quad (x_5 y_5 + x_6 y_6) + (x_7 y_7 + x_8 y_8)$$
$$(x_1 y_1 + x_2 y_2 + x_3 y_3 + x_4 y_4) + (x_5 y_5 + x_6 y_6 + x_7 y_7 + x_8 y_8).$$

This requires some interconnection network to 'bus' the data between the processors. Ideally, a complete interconnection network, when each processor would be directly connected to other processors, would do the job excellently. Unfortunately, it is a very expensive and physically an extremely complex solution for a system with a reasonable number of array processors. Instead, simpler interconnection patterns that are suitable for a broad number of applications are adopted. One simple interconnection scheme is to join each processor to its north, south, east, and west neighbours as shown in Fig. 9.2.4. This pattern is known as a two-dimensional *cyclic-shift interconnection* network. At the grid boundaries the structure is maintained using *wrap-around* connections, e.g. on the east side, PE4 is connected to PE1, PE8 to PE5, etc.

The two-dimensional cyclic-shift interconnection network is useful for summation operations, similar to these outlined in the above example of the scalar product of two vectors. Problems involving partial differential equations which typically require interaction between neighbouring values can also be solved very efficiently on this interconnection network. For instance, in such problems a typical statement is of the form

$$u(i, j) = (1/4)[u(i+1, j) + u(i-1, j) + u(i, j + 1) + u(i, j-1)].$$

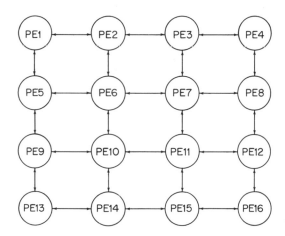

Figure 9.2.4   The cyclic-shift interconnection on a two-dimensional 16-element array processor

Of course, there are many algorithms which require non-cyclic data permutation; for such algorithms the two-dimensional cyclic-shift network is obviously unsuitable.

Of more recent innovative interconnection schemes, one is particularly remarkable; it is known as the *bidirectional perfect shuffle* and at first sight appears artificial. A 'perfect shuffle' of a deck of playing cards is achieved when the deck is split into two equal halves and interlaced in such a manner that cards from alternate halves will succeed one another in a newly compiled deck. Interpreting the idea for the array processors interconnection scheme, the processors are connected in such a manner that the *i*th processor of the unshuffled deck is bidirectionally interconnected to the *i*th processor of the shuffled deck as explained in Fig. 9.2.5(a).

The perfect shuffle pattern is usually used together with a cyclic-shift interconnection network, enhancing the latter into a very powerful network. The perfect shuffle scheme, odd as it seems at first sight, serves extremely well a great variety of parallel algorithms, particularly such important problems as sorting, matrix transposition, and FTs.

Conditional processing

In the example discussed above to find the scalar product of two vectors, at the summation stage of the computation process, only one-half of the processors are involved in shifting their operands to the nearest neighbour for subsequent addition of this operand to the one stored in the neighbour's own memory. This means that when the control processor issues the instruction to 'shift the operand to your neighbour' only some of the PEs—in our example, half of them—should obey and perform this command.

Another example of a similar situation arises in the case of processing of matrices, when the statement of the following form must be executed:

$$\text{if } A(i, j) < 0 \quad \text{then} \quad A(i, j) = 0$$

Each processor can determine if its $A(i, j)$ is less than zero, but not all processors are allowed to carry out the command $A(i, j) = 0$. The problem of this kind is overcome using the facility called *processor-enable mask*; a processor only accepts the instruction from the control processor while its associated bit in the processor-enable mask is set. The control processor must check that the mask is set and reset at the appropriate time. This facility of the enable/disable mask allows a subset of the available processors for an operation to be selected.

Examples of array processors are:

Illiac IV, one-of-a-kind supercomputer which was designed by a group under D. L. Slotnick at the University of Illinois, and built in the late 1960s and early 1970s by the Burroughs Corporation.

Idealized mesh connected computer (MCC), e.g. Illiac IV, (Dekel *et al.*, 1979).

ICL DAP (Parkynson, 1982).

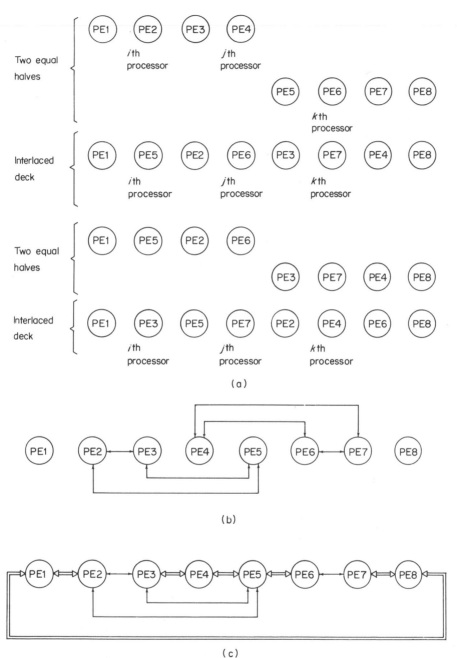

Figure 9.2.5   A cyclic-shift interconnection network augmented with a perfect shuffle scheme. (a) A 'perfect shuffle' mechanism applied to an eight-PE system. It shows that in a 'perfect shuffle' (PE2) should be connected to (PE5) and to (PE3); (PE4) should be connected to (PE6) and to (PE7); (PE6) should be connected to (PE7) and to (PE4), etc. (b) A perfect shuffle interconnection network. (c) A cyclic-shift interconnection network augmented with a perfect shuffle interconnection network

112

Vector and Pipeline Processors

Pipelining implies the provision of a set of functional units, each unit having assigned to it a specific task, e.g. addition, multiplication, logical, etc. In a pipeline processor the data for the operations are defined as vectors (one-dimensional arrays) and the operation is performed as one vector instruction, hence the name—a vector computer. A vector instruction operates on a sequence of operands, a vector. Note that vector processing implies the use of pipelining, but pipelining does not necessarily imply vector processing, e.g. pipelining can be used to carry out a scalar processing, yet in the context of a vector processor, pipelining and vector processing are often used synonymously due to their close relationship. The parallelism of a vector processor is characterized by the vector length.

Pipelining is analogous to an industrial assembly line where the manufactured product moves through a series of stations. Each station executes one step in the manufacturing process and all stations work simultaneously on different units in different stages of completion. Consider a car-manufacturing plant. It may be split up into a number of component stages where each stage carries out a particular operation on the evolving vehicle. Assuming that all stages are individually constrained to take the same time and that each stage requires $t \times$ minutes to complete, the assembly line can move at a rate of one stage per t minutes. Consequently, it requires $t \times$ (the number of stages) minutes for any one particular car to be produced. Now, suppose a vehicle that has just finished the first stage and is moving on to the second stage. The first stage is now free to carry out its particular operation on a new vehicle. At the end of the next t minutes, the second stage can move on to the third stage, the first stage on to the second stage, and the first stage is again free to start on another vehicle. After $t \times$ (the number of stages) minutes the first car has gone through the entire assembly line and is now ready for dispatch. Since there is a vehicle following one stage behind, it will be ready in a further t minutes as well. In fact, now that the assembly line is full, cars will be produced at a rate of one car per t minutes despite the fact that each car requires $t \times$ (the number of stages) minutes to construct.

The vector processing is based on the same principle. An example of a possible floating-point multiplication pipeline is depicted in Fig. 9.2.6.

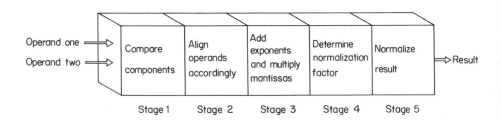

Figure 9.2.6   A simplified pipeline for floating-point multiplication

In the vector processing an instruction set repertoire is used that includes vector instructions. These vector instructions specify a particular operation that is to be carried out on a selected set of operands, i.e. vectors. The control unit issues a vector instruction and the first element(s) of the vector(s) is sent to the appropriate pipe by way of buses. If t units is the time required by one processing stage of the pipe, then after t units of time the second element(s) of the vector(s) is sent to the same pipe using the same buses. Eventually all operands will be transmitted.

Meanwhile, the time between the first operand entering the pipe and the first result appearing at its end is noted and is called the *start-up time*. The start-up time is static for a particular pipe and independent of the length of the vector to be processed. Hence the time to complete the vector instruction is given by

$$\text{start-up time} + t \times (\text{vector length} - 1), \tag{9.2.1}$$

where t is the time taken by one processing stage of the pipe.

From equation (9.2.1) it may be concluded that the vector processing is advantageous for reasonably long vectors but not for 'short' ones, where the start-up time may be comparable to the time of the actual processing of the vector.

Examples of vector computers are:

Cray-1, one of the two commercially available supercomputers of today, which is built by CRAY Research, Inc. (Taylor and Hopkinson, 1982).

Control Data Corporation's (CDC) Cyber 205 roughly competitive with Cray-1 (Purcell, 1982).

CDC 7600, the most successful large-scale scientific computer of the 1970s.

Individual designs may and do combine some or all of the parallel features. For example, a processor array which normally consists of several processors, may have pipelined arithmetic units as its PEs; or in a multi-unit computer, one functional unit may in fact be a processor array.

Comments

The first commercial electronic computer, Univac 1, delivered to the US Bureau of the Census in 1951, was about three times faster than today's home computers and thousands of times more massive. Since Univac the speed of large-scale scientific computers has doubled on the average every two years. The latest supercomputers, the Cray-1 and the Cyber 205, are able to sustain 20 million floating-point operations (megaflops) per second, on average, for a range of typical iterative problems that have a data base of a million or more words.

Although the current performance levels of such machines owe much to the rapid advance of microelectronics, the concept of parallelism in computer architecture has been equally important. Speed of a parallel computer depends on the amount of concurrency inherent in the problem, on how well the inherent concurrency fits the parallel computing capacity designed into the machine. The Cray-1 is the fastest computer for problems dominated by scalar and short-vector operations. The Cyber 205 is the fastest for problems that can be programmed to include long vectors.

Illiac IV was the first computer capable of 20 megaflops and was faster than any model of the Cray or Cyber yet delivered for problems with the largest data base. Since vector computers can execute either scalar or vector instructions only in a single sequential stream, they are called SIMD.

Among the drawbacks of the SIMD-type architectures is the loss of the advantages of the parallel processing whenever a test or a branch instruction occurs, as the system has to wait for the total completion of such instruction before proceeding. The test and branch itself can make no use of the replicated hardware. Another serious problem is that a SIMD computer has a fixed quantity of replicated execution modules or PEs, and hence if the length of the vector which the user wishes to operate on differs from the vector length of the machine, performance suffers and software complexity increases.

Besides, many important large-scale problems cannot be organized into vector forms efficiently. Problems that call for much searching and sorting do not vectorize well. A SIMD machine offers no speed-up for the computation of iteration variables and array subscripts in a scalar problem.

To overcome these difficulties, the concept of a MIMD machine has been discussed for many years. Small experimental MIMD machines have been built at several universities. Availability of mini- and microcomputers in research, industrial, or commercial establishments has prompted the linking of independent processors to form multicomputer systems. Such MIMD systems, however, are restricted by the lack of shared memory between processors. As such they are suitable for executing asynchronous and weakly synchronous parallel programs (Section 9.3). The first superclass MIMD computers are coming into existence currently. An interesting example of such a machine is the heterogeneous element processor (HEP), a multi-processing system developed by Denelcor, Inc. The system's basic uni-processor, the processing-element module (PEM), is enabled to carry out multiple processes by time-sharing the hardware assigned to the control and execution of instructions. Each process is periodically given a chance to execute an instruction, and the logic is pipelined so that several processes are in different phases of execution at any moment of time.

9.3 Parallel Algorithms

A *sequential algorithm* (program) specifies sequential execution of a sequence of statements; its execution is called a process.

A *concurrent algorithm* (program) specifies two or more sequential algorithms (programs) that may be executed simultaneously as parallel processes.

In other words, a process is a collection of sequential control statements which access local and (or) global data and can be executed in parallel with other program units. For example, a holiday booking system that involves processing entries from many terminals has a natural specification as a concurrent program in which each terminal is controlled by its own sequential process.

It is often easier to structure a system as a collection of co-operating sequential processes rather than as a single sequential program, even when processes are not executed simultaneously. A simple operating system can be viewed as three processes:

1. A reader process which reads input into an input buffer;
2. An executer process which reads input images from the input buffer, performs the specified computation (perhaps generating the line image), and stores the results in an output buffer; and
3. A printer process which retrieves line images from the buffer and sends them to a printer.

During the lifetime of a process, different processors of a multiprocessor operating system may be assigned to it on various time intervals, so that, though a part of the program is carried out by one process, its execution may actually be done by many processors.

Asynchronous and Synchronized Algorithms

To ensure that a concurrent algorithm (program) works correctly and effectively, processes interact—communicate, synchronize, and exchange data. The points within the program where the process can communicate with other processes are known as *interaction points*. These interaction points divide a process into stages. At the end of each stage a process may communicate with other processes before starting the next stage. Communication allows execution of one process to influence execution of another.

When processes communicate, synchronization is often necessary. In running a parallel program the time taken by one stage of any of its processes is a random variable. The reasons for this are many: the multiprocessor may consist of processors with different speeds, or a processor, while carrying out a stage of a process may, from time to time, be interrupted by the operating system, or the time of the process may depend on the instances of its input (Kung, 1976).

Thus one can never be sure that an input needed by one process will be produced in time by another process. Synchronization is one way to overcome this problem. In a synchronized algorithm some processes may be blocked from continuing their run until certain stages of other processes are completed. This form of synchronization is called explicit; in a synchronized algorithm the processes wait for inputs whenever necessary.

A different type of synchronization is known as implicit. Implicit synchronization is caused by the contention for shared resources. A parallel program may have a set of global variables accessible to all processes. Upon completion of a stage the process reads global variables, then modifies the values of some global variables, using the results just obtained from the last stage and activates the next stage or terminates itself. In many cases, to ensure logic correctness, the operations on global variables are programmed as *critical sections* (Dijkstra, 1968) and the processes, periodically, may be blocked from entering critical sections which are needed in many programs. Programs exhibiting only implicit synchronization are called asynchronous.

The name 'asynchronous' parallel program suggests that except for global variables, synchronizations are not needed for ensuring that specific inputs are available for processes at various times.

In asynchronous parallel programs, processes never wait for inputs but continue or terminate according to whatever information is currently contained in the global variables. In a synchronized algorithm, on the other hand, a task is decomposed into subtasks which are as much as possible of the same size, so that each subtask is solved by one process of the algorithm. A set of explicit constraints (precedence relations) is then imposed on the ordering in which the stages of different processes are performed. On a completion of some stage, the process may not be activated until another process has finished a certain portion of its program. The precedence relations are enforced by explicit synchronization mechanisms at interaction points.

In the simple operating system with three processes, as described earlier, a shared buffer is used for communication between the reader process and the executer process. These processes must be synchronized so that, for example, the executer process never attempts to read an input image from the input buffer if the buffer is empty.

All the processes that have to synchronize at a given point wait for the slowest among them. If the difference between the speeds of various processes is large, the performance of a synchronized parallel program may be substantially degraded compared with its performance on an 'ideal' multiprocessing system.

Suppose we have a synchronized program with k processes which is run on a multiprocessor system of k identical processors. In an ideal parallel setting this program should run in $1/k$ of the time required in a sequential setting. However, in practice the situation is often different (Kung, 1976).

Let U be the time taken by the program and during the time U let u_p be the total time that p processors are active and $k-p$ processors are idle. Then

$$U = \sum_{p=0}^{k} u_p, \tag{9.3.1}$$

and on a single processor of the system the program can be run in time at most

$$\sum_{p=1}^{k} p u_p . \tag{9.3.2}$$

Hence by using k processors the program is speeded up at most by a factor of

$$S_k = \sum_{p=1}^{k} u_p \Big/ \sum_{p=1}^{k} pu_p \qquad (9.3.3)$$

which is obviously less than the optimal speed-up of $1/k$.

If the u_p are known, S_k may be computed. For instance, if at most $k/3$ processes are active 50 per cent of the time, that is

$$\sum_{p=1}^{k/3} u_p \geq \sum_{p=(k/3)+1}^{k} u_p, \qquad (9.3.4)$$

then

$$S_k = \frac{\displaystyle\sum_{p=1}^{k/3} u_p + \sum_{p=(k/3)+1}^{k} u_p}{\displaystyle\sum_{p=1}^{k/3} pu_p + \sum_{p=(k/3)+1}^{k} pu_p}$$

$$\geq \frac{\displaystyle 2\sum_{p=1}^{k/3} u_p}{\displaystyle (k/3)\sum_{p=1}^{k/3} u_p + k\sum_{p=(k/3)+1}^{k} u_p} \geq \frac{3}{2}\,(1/k). \qquad (9.3.5)$$

Hence in this case the speed-up is at most 67 per cent of what one would expect.

Complexity of Parallel Algorithms

The major resources on which the performance of a sequential algorithm is measured are the time and the memory space. Seeking an analogy to these resources for the parallel algorithms' performance evaluation, we can immediately see that the time remains the major resource in parallel computation, though now it depends not only on the complexity of computation operations of the algorithm but also on the complexity of the overhead operations, such as those created by communication, synchronization, and data exchanges constraints.

Kung (1976) defines the time taken by a parallel program as the elapsed time of the process in the program which finishes last, where the elapsed time of the process is the sum of the following:

(a) basic processing time which is the sum of the times taken by its stages;

(b) blocked time which is the total time that the process is blocked at the end of a stage because it waits for inputs in a synchronized algorithm, or for entering of a critical section in an asynchronous program;

(c) execution time of synchronization overhead, i.e. synchronization house-keeping operations, and implementing critical sections.

In some situations it may happen that the time complexity of a parallel algorithm is actually governed by the time required by the overhead operations rather than the actual computation operations themselves. Consequently, it may be noted that on parallel computers, the goal of minimum execution time is not necessarily synonymous with performing the minimum number of arithmetic operations in the way that it is on a serial computer. For example, when executing on a processor array, bringing together in one processor data from different parts of the memory represents the problem, known as the *routing delays*. This time to *route* data between different processing elements can have an important effect on the efficiency of executing programs. If the time for a typical arithmetic operation is very much longer than the time to pass data between a pair of PEs then the routing delays are relatively unimportant, but if this time is comparable to the routing time, the latter plays an important role in determining the performance of a program and cannot be ignored.

In some situations, a useful performance measure for a parallel program can be the parameter which is quantified as inversely proportional to the CPU time, that is, the time during which the computational units of a computer, arithmetic and logic, used by a program, are engaged during the execution of the program.

On the other hand, the memory factor is no longer an important constraint on a parallel computer. Instead, the hardware size, that is, the number of elements of a parallel machine which are active during a computation, represents a major resource to be taken into a program performance considerations; for an array processor it is the number of processors involved in the computation, and for a vector machine it is the sum of the lengths of the vectors. The idle time of the processors due to interactions with the other processors involved in the program execution—synchronization, communication and data exchanges—is considered a penalty or overhead resulting from the partitioning of the task.

Notation for Parallel Algorithms

Simultaneous operations within a parallel algorithm require special representation conventions. Various notations for specifying parallel or concurrent execution have been proposed (Andrews and Schneider, 1983). For our level of details it suffice to use the ***cobegin-coend*** statement brackets where the degree of parallelism is specified by the range of the index of the **cobegin** statement (Ramakrishnan and Browne, 1983).

For example, the notation

```
cobegin i: ⟨ 0:n−1 ⟩
    operation (A[i], B[i])
coend
```

implies n parallel executions of operation $(A[i], B[i])$ on each of the elements $i = 0, \ldots, n-1$ of A and B.

In relation to this notation, the parallel processors may be assigned a number from 0 to $n-1$, or from 0 to $k-1$, and the processor of index i acts upon the data elements of index i.

Example 9.3.1
Algorithm A
// There are assumed n processors and n elements in each array //

```
begin
    for i:=0 to n−1 do
        cobegin j: ⟨0:n−1⟩
            operation (A[j], B[j], C[j])
        coend
    enddo
end
```

Each execution of the **cobegin–coend** bracket executes simultaneously the operation $(A[j], B[j], C[j])$. This is repeated n times to complete the execution of the algorithm.

Example 9.3.2
Algorithm B
// There are assumed k processors and n and m elements in two arrays respectively, where $m < n$ and $m = ck$, c is an integer. //

```
begin
    for i:=0 to n−1 do
        for j:= 0 to m−1 in steps of k do
            cobegin l: ⟨j:k+j−l⟩
                operation (A[l], B[l], C[l])
            coend
        enddo
    enddo
end
```

An algorithm which can be expressed in the form of Examples 9.3.1 and 9.3.2, would typically show the speed-up proportional to the number of processors if the input and output times, i.e. communications time, is neglected. The latter assumption may be a useful case in some situations of the algorithms' analyses.

In Example 9.3.2 it is easy to see that using k processors the algorithm uses $O(mn/k)$ steps. In general, if n and m are the sizes of two arrays, **A** and **B** respectively, then an algorithm with min(n, m) processors, uses max(n, m) steps.

To express an asynchronously proceeding execution of two or more processes and the return of the results within the main control process, the **fork** and **join** statements will be used.

Example 9.3.3

.

.

.

fork
   Process One:
   Process Two:
   Process Three:
join

.

.

.

9.4 Cascade Partial Sum Method

We shall introduce parallel algorithms for several application areas. However, as the first problem we consider the so-called cascade partial sum parallel computation method, a parallel processing algorithm, which is often used as a part of parallel computation problems in a variety of applications.

Cascade Partial Sum Method

Consider the scalar product of two vectors:

$$c = \sum_{i=1}^{n} a_i b_i .$$

We know that to evaluate this product on a sequential computer in an obvious way (though there are more sophisticted ways of doing it but that is not important here) we must form each product $a_1 b_1, \ldots, a_n b_n$ separately and then add them together. It is also obvious that if we have a parallel computer with n multipliers and sufficient adders we can form all n products simultaneously and then add them together. The immediately evident parallelism of this computation problem is in n independent operations of multiplication and we can capitalize on it by using simultaneously n multipliers. In relation to the process of adding together the individual products, we can intuitively feel that not all adding operations are dependent on each other and we can use this observation to organize the adding procedure as follows:

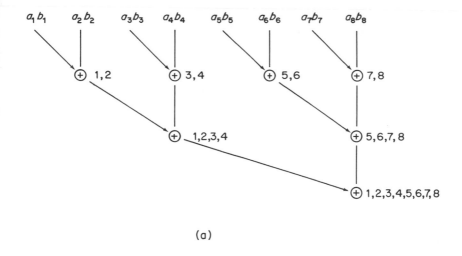

(a)

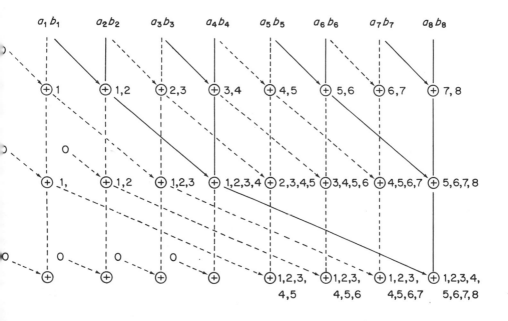

(b)

Figure 9.4.1 (a) Cascade partial sum method for $n = 8$. (b) The routing diagram of the cascade partial sum method implementation on a parallel computer. Zeros are entered as necessary when the vector of accumulators is shifted to the right. The operations shown in broken lines are suppressed if only the total sum $\Sigma a_i b_i$ is required

Add individual products pairwisely;
Add intermediate partial sums pairwisely;
Repeat the process until the total sum is obtained.

Figure 9.4.1 demonstrates the technique in full for $n = 8$. This technique is known as the cascade partial sum method; it is very useful in the algorithms where one tries to make an explicit use of the parallelism of the problem. In the parallel environment the method is often referred to as *cyclic reduction*.

From Fig. 9.4.1 we can see that obtaining the sum of n elements on a parallel computer will require $\log n$ 'addition-time' units as compared with $n-1$ units on a sequential computer. In Fig. 9.4.1 it is also seen that the so-called routing operation is an important characteristic of the performance of a parallel algorithm.

Definition (Hockney and Jesshope, 1981) A unit parallel routing operation is a shift of all elements of an array in parallel to a set of neighbouring processing elements (PEs).

For example, in the case of nearest-neighbour connectivity in a one-dimensional processor array, a unit parallel routing operation is the shift of all elements one PE to the right or one PE to the left, and in a two-dimensional processor array, a unit parallel routing can shift all elements of a two-dimensional data matrix, which is mapped over the processor array, to their nearest-neighbour PEs to the north, south, east, or west.

The number of elements actually shifted after a parallel routing operation is under program control, which means that where appropriate, the shift of some of the elements may be suppressed. In Fig. 9.4.1(b) it is assumed that the horizontal axis represents the relative location of the data in the memory of a pipelined computer, or in the case of a processor array the PE number; the routing operations are shown as solid arrows, pointing to the south-east. When overwriting is undesirable, the variables may be stored in different locations in the same PE of a processor array, or in a different sequence of locations in a pipelined computer.

Summary

The multiprocessor approach introduces three new requirements not encountered before: each problem must be partitioned into tasks, each task must be scheduled for execution on one or more processors, synchronization of control and data flow must be performed during execution. In the following chapters we examine several problems, solutions to which are developed within the support of a multiprocessor environment.

10

The Root-finding Algorithms for a Non-linear Function

The problem is to locate a zero of a continuous function. One classical approach to solve this problem sequentially is to assume an initial root interval, L, that is, an interval which includes the root and is such that the function values have opposite signs at the interval endpoints. Then a nested sequence of approximations to the root is constructed by computing, using an algorithm, a new point inside the interval and evaluating the function at this new point. The root interval is systematically reduced by incorporating the newly computed point and its function value to form one of the new endpoints, on the strength of the test that the function values at the interval endpoints remain at all times of opposite signs. The monotone algorithms based on this approach have global convergence.

Probably the best-known algorithm to compute a new point inside the root interval is to take the midpoint of the interval; this method is known as bisection or, more generally, as binary search, when noting that the approach described can be easily modified to deal with a discrete function, so that, for example, one can use the method for searching an ordered list for a desired item on the list.

10.1 Parallel Zero-searching Algorithms

Synchronized Algorithms

An obvious extension of the bisection method to a search algorithm using k processors is to divide the current root interval into $k+1$ subintervals of equal length and to evaluate the function at each of the k division points. This parallel evaluation is considered as one stage of the root computation process. The computation process is synchronized in the sense that when all k parallel evaluations are complete, a new root interval is computed. The algorithm is depicted in Fig. 10.1.1. We start with a pair of bounds, $x_1^{(0)}$ and $x_2^{(0)}$ for the root x. The iteration may then be described as follows.

123

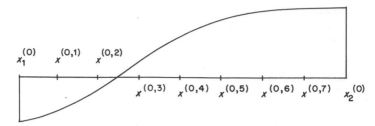

Figure 10.1.1   Parallel bisection method with seven processors

Within the interval $(x_1^{(0)}),\ x_2^{(0)})$ choose k distinct points

$x^{(0,1)},\ x^{(0,2)},\ \ldots,\ x^{(0,k)}.$
cobegin $i:\ \langle l{:}k \rangle$
    Evaluate $f(x^{(0,i)})$
coend
while the root interval greater than prescribed **do**
    Compute an improved interval
    $x_1^{(1)} \leqslant x \leqslant x_2^{(1)}.$
enddo

It is obvious that every parallel evaluation of the function reduces the root interval by a factor of $k+1$, and it has been shown that the order of convergence of the iteration method equals at least the number k of parallel function evaluations.

The major drawback of the synchronized algorithm is that when the times of the k parallel function evaluations differ substantially the algorithm can be very inefficient. An asynchronous zero-searching algorithm is obtained by removing the requirement that the computation process pauses until all k simultaneous function evaluations are complete. We shall outline an asynchronous algorithm on the lines of Kung (1976), whose paper is considered a major contribution to the development of the concepts of asynchronous computations.

An Asynchronous Zero-searching Algorithm

We shall consider an asynchronous algorithm with two processors, in which the choice of the new point is based on the Fibonacci numbers' rule.

Suppose that initially the root interval is divided into three subintervals:

The function evaluation at x_2 and x_3 is started simultaneously by two processors. Suppose now that, without loss of generality, the evaluation at the left point, x_3, finishes first. The assumption here is that the outcome is non-zero, for otherwise a zero is found and we are done. Next, comparison of the values' signs at the left endpoint and at x_3 is executed and the new root interval derived, either as

or as

depending on the sign of the outcome.

If the first case occurs then a new evaluation is carried out at the point x_4 which is defined by

If the second case occurs, the new evaluation is carried out at the point x_5:

In general, in the process of computation one of the two types of computation state can occur, which are denoted by State1(.) and State2(.) and are illustrated in the following graphs:

State1(l)

State2(l)

where $\theta^2 + \theta = 1$, i.e. $\theta = 0.618$ is the reciprocal of the golden ratio, i.e.

$$5/13 + 8/13 = 1, \quad 8/21 + 13/21 = 1, \quad 13/24 + 21/34 = 1, \text{ etc.}$$

State1(l) is the state for which the root interval is of length l and the function is evaluated simultaneously at the point 'o' inside the interval and another point outside the interval (not shown on the graph). Similarly, State2(l) is the state for which the root interval is of length l and the function is evaluated simultaneously at two points, denoted by 'o', both inside the interval. We further deduce that State2(l) is transited after each computation to either

State1($\theta^2 l$),

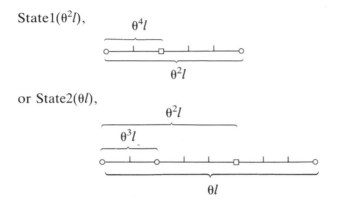

or State2(θl),

This transition is denoted by

$$\text{State2}(l) \rightarrow (\text{State1}(\theta^2 l) \vee \text{State2}(\theta l)) \tag{10.1.1}$$

The corresponding rule for State1(l) is

$$\text{State1}(l) \rightarrow (\text{State1}(\theta^2 l) \vee \text{State1}(\theta l) \vee \text{State2}(l)) \tag{10.1.2}$$

Transition rules (10.1.1) and (10.1.2) completely define the asynchronous parallel algorithm.

Suppose that the algorithm starts from state State2(l). Then, assuming that the function does not vanish at any of the evaluation points, the progress of computation can be represented as a certain transition tree, illustrated in Fig. 10.1.2. The algorithm follows one particular path of the tree, depending upon the input function and the relative computation speeds of the two

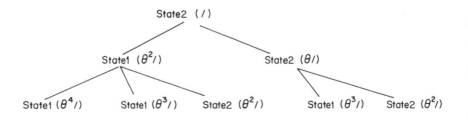

Figure 10.1.2   Transition tree of the parallel bisection method with two processors

processors. Formally the algorithm can be defined as an asynchronous algorithm involving two identical concurrent processes P_i, $i = 1, 2$, which are controlled by the following program (for simplicity it is assumed here that a process is a computation task carried out by one processor of a multiprocessor system):

Process P_i
begin
    while the length of the root interval > 1 **do**
        Compute the position of the next evaluation point '$\square$'   (10.1.3)
        Evaluate the function at the point '$\square$'   (10.1.4)
        Read and update the global variables   (10.1.5)
    enddo
end

The global variables in the program are the positions of the endpoints of the current root interval and the type of the current state.

Step (10.1.3) computes the position of the next evaluation point '$\square$' after examining the global variables. Upon completion of the function evaluation at Step (10.1.4), the global variables are updated at Step (10.1.5). Steps (10.1.3) and (10.1.5) must be programmed within a critical section in order to guarantee the satisfaction of the transition rules (10.1.1) and (10.1.2).

Analysis of the Root-searching Algorithms

We have noted earlier that in the synchronized zero-searching algorithm with k processes, every iteration reduces the length of the root interval by a factor of $k+1$. Hence the algorithm uses $\log_{k+1}L$ iterations to compute the root, where L is the initial length of the root interval. Letting the time needed to evaluate the function at a point in the root interval be a random variable t with mean $\bar{t}$, the expected running time of the algorithm can be shown to be $\lceil\log_{k+1}L\rceil\lambda_k\bar{t}$, where λ_k is the penalty factor for synchronizing k function evaluations. For $k=2$, the expected time estimate for the synchronized algorithm becomes $\lceil\log_3 L\rceil\lambda_2\bar{t}$. This can be compared with the time estimate for the asynchronous two-process algorithm. The readily available transition tree in Fig. 10.1.2 associated with the asynchronous algorithm facilitates the analysis of the algorithm. If N is the number of function evaluations completed by the algorithm and the evaluations are done by two concurrent processes, then the expected time taken by the algorithm is approximately equal to $N\bar{t}/2$ as $N \rightarrow \infty$.

Hence we need to compare $\lceil\log_3 L\rceil\lambda_2$ of the synchronized algorithm with $N/2$ of the asynchronous algorithm, and for this we need to determine the value of N. This value in the worst case is given by the length of the largest path in the transition tree. Analysis of the transition tree carried out by Hayafil and Kung (1975) shows that in the worst case the asynchronous algorithm supersedes the synchronized algorithm with two processes when the penalty factor $\lambda_2 > 1.142$.

The asynchronous algorithm introduced can be generalized to three or more processors. For the case of three processes we can start with the following diagram:

$$L/4 \quad L/4 \quad L/4 \quad L/4$$

$$x_0 \qquad x_2 \qquad x_3 \qquad x_4 \qquad x_1$$

$$L$$

The three processors are activated to evaluate the function at points x_2, x_3, and x_4, chosen as shown in the diagram. As a result of this concurrent function evaluation, without loss of generality, one of the following states will occur:

State1

$$l_1/4$$

$$x_0 \qquad\qquad\qquad x_2$$

$$l_1 = L/4$$

State2

$$l_2/3 \quad l_2/6 \quad l_2/6 \quad l_2/3$$

$$x_2 \qquad x_3 \qquad x_4 \qquad x_1$$

$$l_2 = 3L/4$$

State3

$$l_3/4 \quad l_3/4 \quad l_3/2$$

$$x_0 \qquad x_2 \qquad x_3$$

$$l_3 = L/2$$

where in each case the function is evaluated at point '□'. States 1 and 3 are in fact defining the same pattern

$$l/4 \qquad l/4 \qquad l/4 \qquad l/4$$

$$l$$

while State2 yields the pattern

$$l/3 \quad l/6 \quad l/6 \quad l/3$$

$$l$$

and an asynchronous algorithm with three processes can be fully defined by using the above two patterns.

In general, $\lfloor k/2 \rfloor + 1$ patterns are sufficient for defining an asynchronous algorithm with k processes.

Comments

More recently, another concurrent root-searching algorithm has been proposed by Eriksen and Staunstrup (1983). A real-valued function F, continuous on the interval $[a, b]$, has at least one root in $[a, b]$. Assume that the evaluation of F at any point of $[a, b]$ using a computer is very slow, but it is possible to evaluate F at several points of $[a, b]$ concurrently in essentially the same amount of time that the evaluation at one point takes. The authors have developed a concurrent partitioning algorithm which subdivides the interval until it is known that the root lies in a reduced interval $[a^{(i)}, b^{(i)}]$ where $b^{(i)} - a^{(i)} < \epsilon$ and ϵ is the initially specified maximum error in the value of the root. The algorithm is asynchronous in that it does not wait for all concurrent processes to finish before deciding on a partitioning of the next subinterval. The crucial point of the algorithm is a scheme for maintaining separation of the points at which the function F is being evaluated. The separation is maintained by insisting that two successive points in the partition x_i and x_{i+1} be related by $x_i = \alpha x_{i+1}$, if this is possible, where $0 < \alpha < 1$ is fixed at the beginning. Otherwise a clever approximation scheme is employed. The algorithm's speed is close to the theoretical lower bound on the minimum of the maximum times needed to isolate the root. The authors conjecture a formula for an upper bound on the number of function evaluations which must be completed, but as yet have no proof of the formula.

11

Iterative Parallel Algorithms

The root-finding problem discussed in Chapter 10 is an example of a large class of problems solved by means of the algorithms which repetitively calculate new values of the function or functions using iterative techniques. In iterative techniques a cycle of operations is repeated until the result, or a satisfactory approximation to the result, is obtained. Two types of iterative techniques may be generally recognized: (i) those with guaranteed convergence, however slow it may be (the root-finding bisection method is an example of such a technique); (ii) those that will converge to the solution only if the initial approximation is chosen appropriately, and since the latter choice cannot be always guaranteed the convergence of the corresponding iterative method cannot be guaranteed either.

11.1 Models of Parallel Iterative Methods

A general numerical iterative method can be described by an iteration function

$$\mathbf{x}_{k+1} = \phi(\mathbf{x}_k, \mathbf{x}_{k-1}, \ldots, \mathbf{x}_{k-e+1}), \tag{11.1.1}$$

where, in general, $\mathbf{x}_i$ is the iterative vector produced at the ith iteration and has dimension n; e is the memory of the iterative method, that is, the number of iteration points used in the method (for $e=1$ we have a one-point iterative method).

In order to facilitate parallel computation of equation (11.1.1), one strategy is to exploit the parallelism within the iteration function ϕ. Two forms of decomposition of ϕ are distinguished.

Consider the Jacobi iterative method which solves a linear system of equations:

$$\mathbf{x}_{k+1} = \mathbf{C}\mathbf{x}_k + \mathbf{c} \tag{11.1.2}$$

where $\mathbf{C}$ is the $n \times n$ iteration matrix and $\mathbf{c}$ is a vector of size n. By virtue of the structure of this iterative formula, individual components, $x_{i,k+1}$, of the new iterative vector

130

$$\mathbf{x}_{k+1} = (x_{1,k+1}, x_{2,k+1}, \ldots, x_{n,k+1})$$

can be computed in parallel by more than one process. For example, assume that one would wish to construct a synchronized parallel algorithm consisting of two concurrent processes. Then the most natural approach is to decompose each vector $\mathbf{x}_{k+1}$ into two segments $\mathbf{x}_{k+1}^{(1)}$ and $\mathbf{x}_{k+1}^{(2)}$, each of size $n/2$, and update them by two parallel processes as follows:

$$\begin{bmatrix} \mathbf{x}_{k+1}^{(1)} \\ \mathbf{x}_{k+1}^{(2)} \end{bmatrix} = \begin{bmatrix} \mathbf{C}_{11} & \mathbf{C}_{12} \\ \mathbf{C}_{21} & \mathbf{C}_{22} \end{bmatrix} \begin{bmatrix} \mathbf{x}_k^{(1)} \\ \mathbf{x}_k^{(2)} \end{bmatrix} + \begin{bmatrix} \mathbf{c}^{(1)} \\ \mathbf{c}^{(2)} \end{bmatrix}, \qquad (11.1.3)$$

where $\quad \mathbf{x}_{k+1}^{(1)} = \mathbf{C}_{11}\mathbf{x}_k^{(1)} + \mathbf{C}_{12}\mathbf{x}_k^{(2)} + \mathbf{c}^{(1)}$

and $\quad \mathbf{x}_{k+1}^{(2)} = \mathbf{C}_{21}\mathbf{x}_k^{(2)} + \mathbf{C}_{22}\mathbf{x}_k^{(2)} + \mathbf{c}^{(2)}.$

At each iteration step, each process updates half of the components, the processes are synchronized at the end of each iteration, and the next iteration starts after both processes have finished the updating.

In general, allowing p processors, we obtain a p-process synchronized parallel algorithm for computing equation (11.1.2). We note that the processing time of individual processes (tasks) may vary substantially due to variations in the properties of processing elements, e.g. some instances of the iterates $\mathbf{x}_k$ or some entries of the matrix $\mathbf{C}$ may be null, etc. Thus it is possible that the penalty factor for synchronizing p processes at the end of each iteration is very large, which is an undesirable feature of a synchronized algorithm.

A different approach to decomposing the iteration function ϕ for the purpose of parallel processing of equation (11.1.1) can be illustrated on the example of the Newton–Raphson iterative method for a root-finding problem. Here we have

$$x_{k+1} = \Psi(x_k) = x_k - (f'(x_k))^{-1}f(x_k). \qquad (11.1.4)$$

A natural synchronized iterative algorithm for equation (11.1.4) with two processes is obtained by setting

$$f_1(x_k) = f'(x_k) \quad \text{and} \quad f_2(x_k) = f(x_k) \qquad (11.1.5)$$

and deriving

$$x_{k+1} = \Psi(x_k) = x_k - f_2/f_1. \qquad (11.1.6)$$

At each iteration of equation (11.1.6), $f_2(x_k)$ and $f_1(x_k)$ are computed in parallel, the two processes are synchronized, then the computation for x_{k+1} is started. Generally, f_1 and f_2 may be different, and hence the times needed for their evaluations may differ substantially. For instance, when $f_2 = f$ is a vector of size n, a good approximation to $f_1 = f'$ will need $n+1$ evaluations of f, which means that the process which evaluates $f_2 = f$ wastes a lot of time in waiting at each iteration for the other process to finish the evaluation of f'. This example emphasizes the observation that synchronized iterative algorithms cannot be suitable for those ϕ which do not decompose into mutually independent tasks of the same complexity.

11.2 Model for a Performance Analysis of a Synchronized Iterative Algorithm

Dubois and Briggs (1982) have proposed a model for an analysis of the performance of synchronized iterative algorithms where the function ϕ is decomposable into tasks of approximately equal complexities, i.e. of the kind of iterative function illustrated by the Jacobi method.

The parallel machine assumed in the analysis is a SIMD multiprocessor, set up as a symmetric system of a set of identical processors which operate in a speed-independent manner on shared data. The computation on each processor consists of a random number of instruction executions, where the randomness of the number reflects the variation in the function evaluation techniques.

For simplicity, the implementation of a synchronized iterative process is equated to the efficient implementation of just one iteration step of an iterative algorithm with a given structure. It is assumed that only the concurrent phase of the algorithm is modelled. The input data set includes both the iterates x_k and the parameters defining ϕ, e.g. in the Jacobi method these parameters are the linear system's coefficients.

In order to separate effectively the performance of an iterative algorithm on the assumed architecture, all 'external' fluctuations, e.g. external interrupts and page faults are ignored. In Fig. 11.2.1 the two commonly used multiprocessor systems are diagrammed, displaying particular features assumed in the model, which should ensure 'perfect' performance of the algorithm.

For example, in a loosely coupled system shown in Fig. 11.2.1, the communication memory (CM) is assumed to be double-buffered which ensures that the processor and the direct memory access (DMA) controller can access it concurrently without conflicts. This, in turn, ensures that the time for any given transfer is deterministic. To communicate the information to other processors, a processor initiates a data block transfer through its DMA gate and the interconnection network (IN). The IN is implemented as an interconnection network in its simplest form, as a high-speed bus (HSB) with broadcasting capability. Other implementations of the IN are possible.

To send a message to other processors, a processor stores the message in the CM then initiates the transfer. The DMA controller of the sender monitors the HSB. When it is free, a connection is established on the HSB and is simultaneously read by the $(k-1)$ receivers.

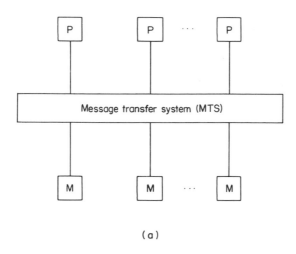

(a)

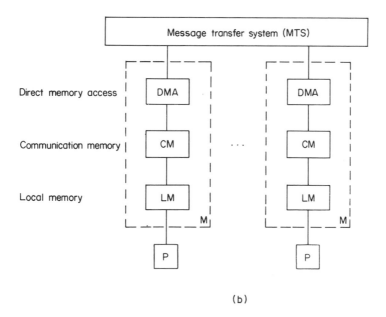

(b)

Figure 11.2.1   Two multiprocessor systems. (a) Tightly coupled multiprocessor; (b) Loosely coupled system with detailed breakdown in the structure of a processor private memory

A computation process in a synchronized iterative algorithm is depicted in Fig. 11.2.2.

For the purpose of an analysis, a set of the machine performance parameters, $p_1, p_2, \ldots, p_m$, is to be extracted from the assumed analytical model and the implementation of a synchronized algorithm is then to be analysed in terms of these parameters. The set of the parameters is also called the parameter space.

Performance of a parallel architecture is in the first place measured by the performance index which is a real function defined on the parameter space of the architecture. Typically, the performance index is chosen to be the average processor utilization time, U. In the case of the model under discussion, as applied to iterative algorithms, where the iteration function is decomposed into approximately equal complexity tasks so that the individual processors have the same average utilization time, we have

$$U = m_0/m_I, \tag{11.2.1}$$

where m_I is the mean iteration time and m_0 the mean time that a procesor is busy processing instructions during each iteration. Hence U is the fraction of time a processor is busy, while $1-U$ is the fraction of time it is idle.

The definition of the utilization time parameter isolates the effect of the architecture on the performance of the iterative algorithm, since (i) in a loosely coupled system, a processor is busy during each processing phase and processor idleness is caused by communication and synchronization at the end of each processing phase, and (ii) in a tightly coupled system, though the communication cost is negligible, U is less than unity because of the machine time wasted in memory conflicts during the processing phases and in waiting for synchronization.

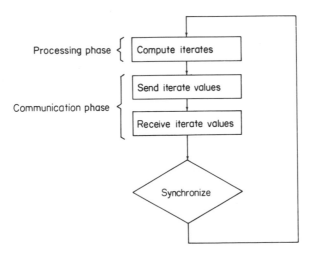

Figure 11.2.2   A flowchart for a process in a synchronized iterative algorithm

Dubois and Briggs (1982) have shown that for the tightly coupled systems the following four performance characteristics of the algorithm can be usefully identified:

p,   the number of processors and thus, in the context of the problem, the decomposition factor of the algorithm;

r,   the probability with which each processor references the shared memory during the processing stage;

m_0,   the mean iteration time;

c_0,   the coefficient of variation in the size of the tasks composing the multiprocessor process; a large c_0 means that some tasks require significantly more computation than others and thus result in larger synchronization waiting times, assuming that:

(a) the processes running on the p processors have the same stochastic properties and are independent;

(b) each processor references the shared memory with a probability r during the processing stage and memory references are uniformly distributed among the memory modules and are independent;

(c) the processes are repeated long enough so that transient effects are negligible;

(d) synchronization does not affect memory conflicts.

The above assumptions are commonly accepted as good approximations to the behaviour of a multiprocessor system.

The study has then shown that for a large p the utilization U is strongly affected by the parameters r and c_0, which leads to the conclusion that more efficient mechanisms for referencing the shared memory should be sought, while for a small p the idle time in some cases may be very substantial.

For loosely coupled systems, it was found that three performance parameters are essential: p, c_0, and t_b, where p and c_0 are defined as in the tightly coupled systems and t_b is the sum of the times t_c and t_r, where t_c is the time required to establish connection on the bus with DMA controllers of the receivers and t_r is the transmission time on the bus, simultaneous with the read time by the receivers. The utilization U is highly sensitive to the ratio 'communication to computation time'; loosely coupled systems are well matched for algorithms with a low 'communication to computation time' ratio. This limits the applicability of loosely coupled architectures for synchronized iterative algorithms.

11.3   Asynchronous Iterative Algorithms

Asynchronous iterative algorithms are naturally developed by removing all the synchronization from the synchronized iterative algorithms. Consider, for example, the Newton–Raphson iteration (11.1.4). In an asynchronous interpretation of the algorithm one can assume that each iteration step updates the three variables $f(x)$, $f'(x)$, and x, rather than x alone. In equation (11.1.4), $f(x_{k-1})$, $f'(x_{k-1})$, and x_k are updated as $f(x_k)$, $f'(x_k)$, and x_{k+1}. Suppose that the evaluation of f' is more expensive than that of f. In this case an asynchronous iterative algorithm consisting of two processes, P1 and P2, can be organized as follows. Three global variables, FX, $F1X$, X, accessible to both processes, contain the current value of $f(x)$, $f'(x)$, and x respectively; $F1X$ is updated by process P1 and FX and X by process P2, in parallel. The two processes are controlled by the following program:

Process P1   **while** condition Stop is not satisfied **do**
                                $F1X := f'(X)$
                         enddo
Process P2   **while** condition Stop is not satisfied **do**
                         begin
                                $FX := f(X)$
                                $X := X - FX/F1X$
                         end
                         enddo

Condition Stop stands for some global criterion for stopping a process, e.g. the absolute value of the difference between two consecutive iterates, x_{k+1} and x_k, must be less than an a priori set tolerance.

As soon as a process finishes updating a global variable, it starts the next updating by using the current values of the relevant variables, without waiting. The computation process is diagrammed in Fig. 11.3.1. In this setting, the asynchronous iterative formula, in general, would be:

$$x_{k+1} = x_k - f(x_k)/f'(x_j), \qquad (11.3.1)$$

where $j \leqslant k$.

Thus the asynchronous algorithm generates a sequence of iterates which is different from that generated by the synchronized iterative algorithm (or by the sequential algorithm).

For a general iterative process (11.1.1) an asynchronous iterative algorithm can be developed by first identifying some variables $V1, \ldots, Vm$ such that each iterative step can be regarded as computing the new values of the Vj's from their old values. For the Newton–Raphson iteration (11.1.4) the choice $(V1, V2, V3) = (FX, F1X, X) = (f(x), f'(x), x)$ means that the updating of each Vj constitutes a significant portion of the work involved in one iteration. This principle is recommended to adhere in general when constructing an asynchronous iterative algorithm.

Another point that may be noted about the asynchronous Newton–Raphson iteration is that the two processes of the algorithm are identified by

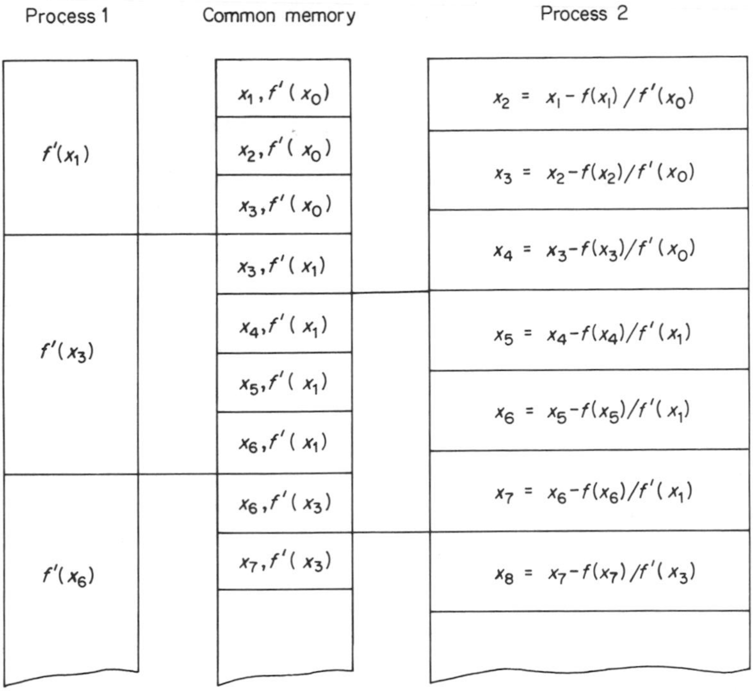

Process 1	Common memory	Process 2
	$x_1, f'(x_0)$	$x_2 = x_1 - f(x_1)/f'(x_0)$
$f'(x_1)$	$x_2, f'(x_0)$	$x_3 = x_2 - f(x_2)/f'(x_0)$
	$x_3, f'(x_0)$	
	$x_3, f'(x_1)$	$x_4 = x_3 - f(x_3)/f'(x_0)$
$f'(x_3)$	$x_4, f'(x_1)$	$x_5 = x_4 - f(x_4)/f'(x_1)$
	$x_5, f'(x_1)$	$x_6 = x_5 - f(x_5)/f'(x_1)$
	$x_6, f'(x_1)$	
	$x_6, f'(x_3)$	$x_7 = x_6 - f(x_6)/f'(x_1)$
$f'(x_6)$	$x_7, f'(x_3)$	$x_8 = x_7 - f(x_7)/f'(x_3)$

Figure 11.3.1 A diagram of an asynchronous Newton–Raphson method with two processors for solving a single non-linear equation

specific permutation on the set $(V1, V2, V3)$, namely $(V1, V3) = (FX, X)$ and $(V2) = (F1X)$.

In general, one can deduce that a process can be specified by a permutation on the set of the global variables $(V1, \ldots, Vm)$ in the following sense: the process updates the Vj's in the relevant subset sequentially according to the sequence which defines the permutation. Thus, given a permutation on a set of global variables $(V1, \ldots, Vm)$, an asynchronous iterative algorithm is a collection of the processes which work asynchronously and have the property that each Vj appears in at least one of the subsets associated with the processes. This condition guarantees that every Vj is taken care of by at least one process.

For example, in an asynchronous algorithm (11.1.2), Vj's may be chosen as segments of equal size of the components in a vector iterate, and concurrent processes which update the Vj's asynchronously can be defined in the way that ensures that every Vj is computed in at least one of these concurrent processes.

We shall now consider asynchronous Newton-like algorithms for solving several non-linear equations.

11.4 Parallel Algorithms for Solving Systems of Non-linear Equations

Consider the following set of non-linear algebraic equations:

$$f_1(x_1, \ldots, x_n) = 0,$$
$$f_2(x_1, \ldots, x_n) = 0,$$
$$\ldots \qquad\qquad\qquad (11.4.1)$$
$$\ldots$$
$$\ldots$$
$$f_n(x_1, \ldots, x_n) = 0,$$

where functions f_j are assumed to be continuously differentiable.

The Newton–Raphson method for locating a solution to equation (11.4.1) is expressed in a vector form by

$$\mathbf{x}_{k+1} = \mathbf{x}_k - \mathbf{J}(\mathbf{x}_k)^{-1}\mathbf{F}(\mathbf{x}_k), \quad \mathbf{x}_0 \text{ is arbitrary,} \qquad (11.4.2)$$

where $\mathbf{J}(\mathbf{x})^{-1}$ is the inverse of the Jacobian matrix of $\mathbf{F}(\mathbf{x})$. (The Jacobian matrix contains the $n \times n$ partial derivatives of $\mathbf{F}$ with respect to $\mathbf{x}$, where $\mathbf{J}_{ij} = \partial F_i/\partial x_j$.)

The method's convergence is of order two (Ortega and Rheinboldt, 1970); that is, if

$$\mathbf{x}_k \to \mathbf{x}, \text{ the solution-vector, as } k \to \infty,$$

then

$$\|\mathbf{x}_k - \mathbf{x}\| \leq K\theta^{2^k} \quad \text{for all } k, \text{ for some } K > 0, \text{ and } \theta \in (0, 1).$$

When the direct evaluation of the Jacobian matrix $\mathbf{J}(\mathbf{x}_k)$ is too complex and is preferably to be avoided, the following finite-difference approximation, $\mathbf{J}_k$, to $\mathbf{J}(\mathbf{x}_k)$ may be used, yielding the method known as the discrete Newton–Raphson method and due to Shamanskii (1967).

Let $\mathbf{c}_j$ be the jth column of the $n \times n$ identity matrix, $\mathbf{I}$, and let the jth column $\mathbf{J}_k\mathbf{c}_j$ of the matrix $\mathbf{J}_k$ be defined

$$\mathbf{J}_k\mathbf{c}_j = [\mathbf{F}(\mathbf{x}_k + \epsilon_k\mathbf{c}_j) - \mathbf{F}(\mathbf{x}_k)]/\epsilon_k, \quad 1 \leq j \leq n, \qquad (11.4.3)$$

where ϵ_k satisfies $0 < \epsilon_k \leq \|\mathbf{F}(\mathbf{x}_k)\|$.

Given a solution estimate x_k, the discrete Newton–Raphson method generates the next estimate x_{k+1} by

$$x_{k+1} = x_k - J_k^{-1} F(x_k). \tag{11.4.4}$$

The order of convergence for algorithm (11.4.4) is two, the same as that for the Newton–Raphson method. The following algorithm is based on this method.

Algorithm newton (root) (11.4.5)

Initialize rootestimate
Initialize Stop
while not Stop **do**
    Compute **J** through (11.4.3)
    Compute J^{-1}
    rootestimate := rootestimate $- J^{-1} F$(rootestimate)
    Compute new Stop
enddo

In each of the two methods either the Jacobian matrix or $(n+1)$ gradients, $F(x_k)$, $F(x_k+\epsilon_k c_j)$, $j=1, 2, \ldots, n$, are required to be evaluated at each iteration, that is, much computation time is spent evaluating the current Jacobian or its finite-difference approximation.

As a means of speeding up the progress, one can think of spreading out the computational work of the Jacobian evaluation over several iterations. Two strategies have emerged following these lines of thought.

One strategy, following Mukai (1979), may be called partial updating. It involves the updating of only one or at most k, $k < n$, columns of the Jacobian at a time and retaining the remaining columns from the previous iteration (Wolfe, 1959; Barnes, 1965; Polak, 1974).

The other strategy may be called deferred updating and can be seen in methods proposed by Traub (1964), Shamanskii (1967), Brent (1973), and Mukai (1979). In the deferred updating approach, the Jacobian is only occasionally re-evaluated. Given a solution estimate x_k, the next estimate x_{k+1} is computed as a result of s Newton–Raphson iterations with the same Jacobian or its finite-difference approximation. The following algorithm is based on the Shamanskii method.

Algorithm shamanskii (rootestimate, s) (11.4.6)

Initialize rootestimate
Initialize Stop
while not Stop **do**
   Compute $\mathbf{J}$ through algorithm (11.4.3)
   Compute $\mathbf{J}^{-1}$
   $i := 0$
   while not Stop **and** $i < s$ **do**
     $i := i + 1$
     rootestimate := rootestimate $- \mathbf{J}^{-1}\mathbf{F}$ (rootestimate)
     Compute new Stop
   enddo
enddo

Both approaches originally evolved as a result of seeking to speed up the sequential Newton–Raphson method, but due to their inherently parallel set-up, lend themselves beautifully to parallel implementation. An asynchronous algorithm based on the finite-difference approximation, $\mathbf{J}$, to the Jacobian can be formulated as follows.

Algorithm asynewton (rootestimate) (11.4.7)

Initialize rootestimate
Initialize Stop
if not Stop **then**
   Compute $\mathbf{J}$ through algorithm (11.4.3)
   Compute $\mathbf{J}^{-1}$
   rootestimate := rootestimate $- \mathbf{J}^{-1}\mathbf{F}$ (rootestimate)
   Compute new Stop
   fork
   Process One: **while not** Stop **do**
                Compute $\mathbf{J}$ through algorithm (11.4.3)
                Compute $\mathbf{J}^{-1}$
            enddo
   Process Two: **while not** Stop **do**
                rootestimate := rootestimate
                   $- \mathbf{J}^{-1}\mathbf{F}$ (rootestimate)
                Compute new Stop
            enddo
   join
endif

In the algorithm (11.4.7) the iterations with a fixed Jacobian are carried out only as long as the time required to compute a new Jacobian. The algorithm is fully asynchronous and can be implemented on a MIMD machine.

In an analogous parallel semi-asynchronous Newton–Raphson scheme which uses a fixed number of iterations, s, with each Jacobian, care should be taken to set the number s such that the time taken by the corresponding 'block' of iterations is equal to or greater than the time required to compute an updated Jacobian.

Efficiency Analysis of Iterative Processes

In order to compare iterative processes of differing speeds, Brent (1973) has defined the efficiency of an iterative process by

$$E = (\log\rho)/w, \tag{11.4.8}$$

where

$$\rho = \lim_{i\to\infty} \inf(-\log \|\mathbf{x}_i - \mathbf{x}\|)^{1/i} \quad (> 1)$$

is the order of convergence of the process and

$$w = \lim_{i\to\infty} w_i \quad (> 0)$$

is the asymptotic bound on the amount of work required to compute $\mathbf{x}_i$ from $\mathbf{x}_{i-1}$ and other previous iterations ($\mathbf{x}$ denotes the exact root).

To compare parallel algorithms, where time is a more appropriate measure than work, Mukai (1979) has modified algorithm (11.4.8) and defined the time efficiency of an iterative method by

$$E = (\log\rho)/t \tag{11.4.9}$$

where

$$t = \lim_{i\to\infty} t_i \quad (> 0)$$

is the asymptotic bound on the amount of time required to compute $\mathbf{x}_i$ from $\mathbf{x}_{i-1}$ and other previous iterations. With this definition, a method with time efficiency E requires E'/E times as much time as a method with time efficiency E' to reduce $\|\mathbf{x}_i - \mathbf{x}\|$ to a very small positive tolerance. It must be noted, however, that the method with the higher efficiency is not necessarily the method with the higher order of convergence.

Efficiency Analysis of Newton–Raphson Algorithm

For algorithm (11.4.5), Traub (1964) has shown that the order of convergence is two. If t_J is the time required to compute $\mathbf{J}$ (through formula (11.4.3)) and $\mathbf{J}^{-1}$, and t_x is the time required to compute $\mathbf{x}$ (through formula (11.4.4)) and test for convergence, then the time efficiency of the algorithm is

$$E_N = (\log 2)/(t_J + t_x) \tag{11.4.10}$$

For algorithm (11.4.6), Shamanskii (1967) has shown that the order of convergence is at least $(s + 1)$—Traub (1964) has obtained the same result for an algorithm that uses the exact Jacobian matrix—thus the time efficiency of the algorithm is at least

$$E_{Sh} = (\log (s+1))/(t_J + st_x).$$

If s^* is the s that maximizes E_{Sh}, then the optimal time efficiency of the algorithm is at least

$$E_{Sh} = (\log (s^*+1))/(t_J + s^*t_x). \tag{11.4.11}$$

Given an estimate for the root, $\mathbf{x}_i$, Process Two of algorithm (11.4.7) generates the next estimate as a result of one or more Newton–Raphson iterations with the approximate Jacobian matrix $\mathbf{J}_j$, where $j < i$. For this algorithm, t is approximately the time Process Two takes to complete its computations. Thus the time efficiency of the algorithm is

$$E_{AsN} = (\log\rho_{AsN})/t_J. \tag{11.4.12}$$

The Relationship Between t_J and t_x

If it is assumed that arithmetic operations and evaluations of $\mathbf{F}(\mathbf{x})$ dominate the computing time, then t_J and t_x can be expressed in terms of three constants c_1, c_2, and c_3, where $c_1 n^2$ is the time required to compute $\mathbf{F}(\mathbf{x})$, c_2 is the time required to perform an addition (or subtraction), and c_3 is the time required to perform a multiplication (or division). Thus from the algorithms we have

$$t_J = (c_1 + c_2 + c_3)n^3 + (c_1 + c_3)n^2 + c_2n, \tag{11.4.13}$$

$$t_x = (c_1 + c_2 + c_3)n^2 + 2c_2n, \qquad (11.4.14)$$

and thus the relationship between t_J and t_x is

$$\lim_{n \to \infty} t_J/t_x = n. \qquad (11.4.15)$$

This result was obtained by Levin (1984) and further supported by experimental work.

The Order of Convergence of the Asynchronous Algorithm

If t_x is set to unity in equation (11.4.15), the time efficiencies of algorithms (11.4.5), (11.4.6), and (11.4.7), given in equations (11.4.10), (11.4.11), and (11.4.12) respectively, are then given by

$$
\begin{aligned}
E_N &= (\log 2)/(n + 1), & (11.4.16) \\
E_{Sh} &= (\log(s^* + 1))/(n + s^*), & (11.4.17) \\
E_{AsN} &= (\log \rho_{AsN})/n. & (11.4.18)
\end{aligned}
$$

Brent (1973) has obtained the same expressions for E_N and E_{Sh} as given by equations (11.4.16) and (11.4.17) respectively, although his analysis is based on a different assumption, and shows that for all $n \geq 1$, algorithm (11.4.6) is more efficient than algorithm (11.4.5). For $n = 1$ (a single non-linear equation) the difference is slight, but the difference is appreciable for $n > 1$ (a set of non-linear equations).

The greater efficiency of the Shamanskii algorithm (11.4.6) over the Newton algorithm (11.4.5) was observed in the experimental results of Levin (1984). The results have further shown that the efficiencies of algorithms (11.4.6) and (11.4.7), for $1 \leq n \leq 10$, are approximately equal. Thus, from equations (11.4.17) and (11.4.18), the order of convergence of the asynchronous Newton algorithm (11.4.7) is at least

$$\rho_{AsN} = (s^* + 1)^{n/(n+s^*)}, \qquad (11.4.19)$$

whether this approximation holds for $n > 10$ remains to be investigated.

In a theoretical analysis of an asynchronous algorithm similar to algorithm (11.4.7), Baudet (1978) has shown that, if t_J/t_x is close to n, then its order of convergence is at least λ_n, where λ_n is the largest root of the equation

$$z^3 - z^2 - (n - 1)z - 1 = 0.$$

Table 11.4.1 shows that the values of ρ_{AsN} and λ_n for $1 \leq n \leq 10$ do not differ greatly.

Table 11.4.1 An estimated order of convergence of an asynchronous Newton algorithm for solving a set of n non-linear equations

n	s	ρ_{AsN}	λ_n
1	2	1.44	1.47
2	3	1.74	1.84
3	3	2.00	2.15
4	4	2.24	2.41
5	5	2.45	2.65
6	5	2.66	2.87
7	6	2.85	3.06
8	6	3.04	3.25
9	7	3.22	3.42
10	7	3.40	3.59

Comments

The asynchronous Newton algorithm (11.4.7) was implemented by Levin (1984) by simulation on a sequential computer and so the results reported will have to be checked against those obtained by an implementation of the algorithm in a genuine MIMD environment.

Also as Kung (1976) has observed, for solving a given problem it is almost always possible to construct a large number of asynchronous algorithms. Thus, constructing a different asynchronous algorithm to that of algorithm (11.4.7), might also affect the algorithm's performance.

Epilogue

Both types of concurrent iterative algorithms have serious drawbacks; the synchronized algorithms may have processes blocked for a long time due to synchronization restrictions while the asynchronous algorithms are very difficult to analyse. One can then look at the ways in which the positive features of both types of iterative algorithms can be usefully compromised without being limited by the drawbacks of either of the two types of algorithms. This leads to the so-called semi-synchronized or semi-asynchronous iterative algorithms. For further results on these see, for example, Kung (1976), Chazan and Miranker (1969).

12

Synchronized Matrix Multiplication Algorithms

Multiplication of matrices constitutes one of the principal computation stages in solving many diverse problems. On serial computers, the algorithms for solving these problems require large amounts of storage and time.

On the other hand, matrix multiplication is one of the classical problems particularly suited for parallel processing. The $n \times n$ product of **A** of two $n \times n$ matrices **B** and **C** is given by

$$a(i, j) = \sum_{k=1}^{n} b(i, k) \, c(k, j) \quad \text{for } i = 1, \ldots, n \qquad (12.1)$$
$$j = 1, \ldots, n$$

It is easily observed that the $a(i, j)$'s are computed using three loops on the indices: an inner, middle, and outer. The inner loop contains the expression

$$a(i, j) := a(i, j) + b(i, k)c(k, j),$$

and by varying the loop indices, i, j, k, different matrix product algorithms can be derived.

12.1 Matrix Multiplication Algorithms

Six permutations are possible for arranging the three loop indices. Each of these permutations differs in how the matrix elements are accessed, i.e. by row or by column, as a scalar, vector, or matrix. Each permutation has a different memory access pattern and it is this pattern that affects a particular algorithm's performance on a specified parallel machine. The permutations are given in Fig. 12.1.1 (Dongarra, 1983). Within each of the algorithms, the initialization of the array is placed in natural locations.

All the algorithms shown in Fig. 12.1.1 perform the same operation in exactly the same order, but their performance when implemented on a parallel processor can vary greatly because of the way information is accessed. What has been done in Fig. 12.1.1 is simply 'interchange of loops'. This rearrangement can dramatically affect performance, depending on the type of parallel processor.

146

```
for i:=1 to n do
    for j:=1 to n do
        a(i, j):=0
        for k:=1 to n do
            a(i, j):=
                a(i, j)+b(i, k)×c(k, j)
        enddo
    enddo
enddo
```
(a) Form *ijk*

```
for j:=1 to n do
    for i:=1 to n do
        a(i, j):=0
        for k:=1 to n do
            a(i, j):=a(i, j)+
                b(i, k)×c(k, j)
        enddo
    enddo
enddo
```
(b) Form *jik*

```
for i:=1 to n do
    for j:=1 to n do
        a(i, j):=0
    enddo
enddo
for k:=1 to n do
    for i:=1 to n do
        for j:=1 to n do
            a(i, j):=
                c(i, j)+ b(i, j)×c(k, j)
        enddo
    enddo
enddo
```
(c) Form *kij*

```
for j:=1 to n do
    for i:=1 to n do
        a(i, j):=0
    enddo
enddo
for k:=1 to n do
    for j:=1 to n do
        for i:=1 to n do
            a(i, j):=
                a(i, j)+b(i, j)×c(k, j)
        enddo
    enddo
enddo
```
(d) Form *kji*

```
for i:=1 to n do
    for j:=1 to n do
        a(i, j):=0
    enddo
enddo
for k:=1 to n do
    for j:=1 to n do
        a(i, j):=
            a(i, j)+b(i, k)×c(k, j)
    enddo
enddo
enddo
```
(e) Form *ikj*

```
for j:=1 to n do
    for i:=1 to n do
        a(i, j):=0
    enddo
enddo
for k:=1 to n do
    for i:=1 to n do
        a(i, j):=
            a(i, j)+b(i, k)×c(k, j)
    enddo
enddo
enddo
```
(f) Form *jki*

Figure 12.1.1 Six permutations for arranging the three loop indices in the matrix multiplication

We shall study the design and analysis of the parallel matrix multiplication algorithms with reference to three different synchronous configurations, a conceptual MCC, the Cray-1-like structure, and the ICL DAP-like structure.

In the algorithms of the form *ijk* and *jik* the inner loop is performing an inner product calculation. Their transformation into a parallel algorithm with the optimum time complexity is as follows:

Perform all of the required scalar products in parallel and create a three-dimensional structure in which to store the results.

Add all of its components together, thus eliminating the third dimension of the structure and producing a two-dimensional structure in which the product matrix is recorded; this step requires $O(\log_2 n)$ time.

Algorithm matrix prodOne (b, c: *matrices*)

```
cobegin i: ⟨1:n⟩
   cobegin j: ⟨1:n⟩
      cobegin k: ⟨1:n⟩
         temp(i, j, k) := b(i, k) × c(k, j)
      coend
      a(i, j) := innersum(temp(i, j, 1:n))
   coend
coend
```

Innersum is a function whose argument is a vector and whose result is the sum of the components of the vector. The function can be implemented by an algorithm which uses recursively the cascade partial sum method, i.e. adds the components of two halves of a vector together, producing a vector whose length is half that of its argument, until one element remains. The best theoretical time complexity for this matrix multiplication algorithm is $O(\log_2 n)$ and the algorithm requires $O(n^3)$ space.

Forms *kij* and *kji* can be transformed into a parallel algorithm which avoids a three-dimensional structure by means of replacing some of the parallel operations by sequential ones, i.e. all inner products associated with the product matrix are accumulated sequentially.

Compute all first terms of the inner products.

Compute all second terms of inner products and accumulate (add) the results with the first terms.

Continue the process until the complete inner products are computed.

Algorithm matrix prodTwo (b, c: matrices)

```
cobegin i: ⟨1:n⟩
  cobegin j: ⟨1:n⟩
    temp(i, j) := 0
    for k:=1 to n do
      temp(i, j) := temp(i, j) + b(i, k) × c(k, j)
    enddo
    a(i, j) := temp(i, j)
  coend
coend
```

The algorithm requires $O(n^2)$ space and its time function is of $O(n)$.

Forms *ikj* and *jki* can be transformed into a parallel algorithm which makes use of the fast copying (broadcasting) facility of the parallel machine. The algorithm is as follows:

Select one vector from each of the two matrices. Duplicate both *n* times to form new matrices.

Multiply the corresponding components of these newly constructed matrices.

Note. The vectors are chosen in such a way that the (i, j)th element of the component product of the matrices forms one term of the inner product associated with the corresponding element of the product matrix, e.g.

$$
\begin{matrix}
a_{11} & a_{11} & a_{11} \\
a_{21} & a_{21} & a_{21} \\
a_{31} & a_{31} & a_{31}
\end{matrix}
\qquad
\begin{matrix}
b_{11} & b_{12} & b_{13} \\
b_{11} & b_{12} & b_{13} \\
b_{11} & b_{12} & b_{13}
\end{matrix}
$$

gives the product

$$
\begin{matrix}
a_{11}b_{11} & a_{11}b_{12} & a_{11}b_{13} \\
a_{21}b_{11} & a_{21}b_{12} & a_{21}b_{13} \\
a_{31}b_{11} & a_{31}b_{12} & a_{31}b_{13}
\end{matrix}
$$

Repeat the process, adding each new component of the inner products on to the previously accumulated partial sums until all of the elements of the product matrix have been computed, e.g.

$$
\begin{matrix}
a_{12} & a_{12} & a_{12} \\
a_{22} & a_{22} & a_{22} \\
a_{22} & a_{32} & a_{33}
\end{matrix}
\qquad
\begin{matrix}
b_{21} & b_{22} & b_{23} \\
b_{21} & b_{22} & b_{23} \\
b_{21} & b_{22} & b_{23}
\end{matrix}
$$

and the sum accumulated so far is

$$
\begin{matrix}
a_{11}b_{11}+a_{12}b_{21} & a_{11}b_{12}+a_{12}b_{22} & a_{11}b_{13}+a_{12}b_{23} \\
a_{21}b_{11}+a_{22}b_{21} & a_{21}b_{12}+a_{32}b_{22} & a_{21}b_{13}+a_{23}b_{23} \\
a_{31}b_{11}+a_{33}b_{21} & a_{31}b_{12}+a_{33}b_{23} & a_{31}b_{13}+a_{33}b_{23}
\end{matrix}
$$

Algorithm matrix prodThree (b, c: *matrices*)

```
cobegin i: ⟨1:n⟩
  cobegin j: ⟨1:n⟩
    temp(i, j) := 0
    for k:=1 to n do
      matrixofb(i, j) := b(i, k)
      matrixofc(i, j) := c(k, j)
      temp(i, j) := temp(i, j)
        + matrixofb(i, j)×matrixofc(i, j)
    enddo
    a(i, j) := temp(i, j)
  coend
coend
```

This is again an algorithm of $O(n)$ time and of $O(n^2)$ space complexities.

A radically different approach to matrix multiplication has been proposed in Cannon's algorithm. Its time function at best is of $O(n)$ and the algorithm requires $O(n^2)$ space. Two distinguishable stages define the algorithm.

Matrices **B** and **C** are aligned in such a way that each component of the elementwise product $b(i, j)c(i, j)$ gives one of the required n terms in the computation of the inner product associated with the corresponding element of the product matrix. This alignment is achieved by cyclically shifting (which requires a special operator, *Rotate*) the components of each row, i, in **B**, $(i-1)$ places to the west and the components of each column, j, in **C**, $(j-1)$ places to the north. We have

$$a(i, j) := a(i, j) + b(i, k)b(k, j) .$$

The loop on k is replaced by:

Rotation

$$b_{ij} = b_{i,j \; Rot \; 1}, \qquad \text{if } i > k$$
$$c_{ij} = c_{i Rot \; 1,j} , \qquad \text{if } j > k,$$

e.g. for $k = 1$

b_{11}	b_{12}	b_{13}		c_{11}	c_{22}	c_{33}
b_{22}	b_{23}	b_{21}		c_{21}	c_{32}	c_{13}
b_{33}	b_{31}	b_{32}		c_{31}	c_{12}	c_{23}

=rows to the west=	=columns to the north=
ith row is shifted	jth column is shifted
$(i-1)$ places to the west	$(j-1)$ places to the north.

The rows of **B** are cyclically shifted one place west, the columns of **C** are cyclically shifted one place north, and the product $b(i, j)c(i, j)$ is again evaluated to give another term in each inner product required in the formation of the result matrix. This term is then added on to the previously accumulated partial sum.

Repeat this step until all of the required inner products have been computed.

Algorithm alignment

```
for k := 1 to n do
    cobegin i: ⟨1:n⟩
        cobegin j: ⟨1:n⟩
            if i>k then b(i, j) := b(i, j Rot 1)
            if j>k then c(i, j) := c(i, Rot 1, j)
        coend
    coend
enddo
```

Algorithm form products

```
cobegin i: ⟨1:n⟩
    cobegin j: ⟨1:n⟩
        temp(i, j) := b(i, j)×c(i, j)
        for k:=1 to n do
        begin
            b(i, j) := b(i, j Rot 1)
            c(i, j) := c(i Rot 1, j)
            temp(i, j) := temp(i, j) + b(i, j)×c(i, j)
        enddo
        a(i, j) := temp(i, j)
    coend
coend
```

12.2 Parallel Hardware Considerations for the Matrix Product Algorithms
(Clint *et al.*, 1983)

Each of the four matrix product algorithms discussed uses particular parallel processing features. It may happen that for a particular machine it is not possible to implement efficiently some of the basic features of an algorithm. It may result in an inefficient machine code.

Consider the MCC environment. We recall that the MCC is an array processor consisting of a number of identical processors each constrained to execute the same instruction simultaneously on its own local data. Assume that such a machine has p^2 processors which are arranged in a square grid with cyclic-shift interconnection (see Fig. 9.2.4) and are synchronized. An enable/ disable mask is available to select a subset of the processors for an operation.

The first three algorithms for matrix multiplication are all inefficient on the MCC: a three-dimensional network required by the fast algorithm is not available on the MCC, thus making its execution impractical; too much data movement required in the second algorithm to make its execution efficient; and the third algorithm requires a rapid means of duplicating a vector n times over the processors (broadcasting facility), which is not available on the MCC. Cannon's algorithm is recommended for use on the MCC, but even here certain further considerations have to be explored, see, for example, Dekel *et al.* (1979).

Another parallel processor, the DAP, is a bit processor, its parallel processing capacity being provided by a 64×64 square grid of single-bit PEs, each with its own local data memory. A selection of active processors is achieved by masking part of the grid. The interconnection network consists of (i) a cyclic-shift connection on the grid, and (ii) broadcasting facilities via some globally accessible registers. Thus the DAP can be visualized as an extension of the MCC model which includes an additional mechanism— broadcasting—for data transfer. On the DAP, matrix multiplication may be performed by selecting one vector from each of the two matrices to be multiplied, broadcasting them over the processor mesh, and multiplying the corresponding components of the new matrices. This means that the third matrix algorithm can be efficiently implemented on the DAP. The broadcasting facility alleviates the problem of data movement. The implementation of the Cannon algorithm is the same as on the MCC, but it does not use the broadcast facilities provided. However, the first and the second matrix multiplication algorithms still remain inefficient for implementing on the DAP since they require too much data movement in the context of the DAP communication network. Particular consideration should also be given to the problems where the problem size differs from the DAP 64×64 PE structure.

The Cray-1 processor is a parallel machine equipped with a set of vector and scalar functional units, with a specific task, e.g. add, multiply, etc. assigned to each of the units. Vector operations are performed in a vector register configuration with concurrency provided by pipelining. Only vectors of length 64 or less can be processed. The longer vectors have to be split into segments of length 64 and, possibly, one remaining segment of length less than 64. All stages in the pipeline are synchronized. The ways in which the Cray-1 perform the active operations selection on a vector are as follows. If the increments between the indices of the vector components to be selected are constant (the regular selection pattern) then the required subset of components is extracted from the original vector and processed. If the selection pattern is irregular then the entire vector is processed first and thereafter the relevant components only are copied into memory by means of a logical mask. Therefore, the time required to process an irregular subset of a vector is equivalent to the time to process the entire vector. This means that sequential scalar processing may be preferable in the cases where the irregularly selected subset is small compared with the length of the entire vector.

Another advanced form of parallelism implemented on the Cray-1 is called chaining. Chaining is used to overlap functional unit execution and means that several of the vector functional units may be in use concurrently, each performing a separate operation and the output from one functional unit is directly passed as input to another different functional unit. Chaining is useful when a number of different operations are applied to one set of data. On a vector computer it is important to distinguish how data are handled with respect to the vector registers and memory. The Cray-1 is limited in the sense that there is only one path between memory and the vector register. This creates a bottleneck if a program loads a vector from memory, performs some arithmetic operations, and then stores the results. While the load and arithmetic can proceed simultaneously as a chained operation, the store is not started until that chained operation is fully completed.

In view of the fact that the hardware on the Cray-1 is constrained to process vectors of a maximum length 64, it is convenient to distinguish between the parallel matrix-multiplication operation on (i) matrices of size 64 or less, and (ii) matrices of size larger than 64.

There is also another reason for distinguishing between the 'small' and 'large' matrix-multiplication operations on the Cray-1, and, in general, on the Cray-1-like architectures. This reason is the way in which the matrix symmetry is treated on a pipeline machine.

Suppose that one wishes to compute the off-diagonal elements of a symmetric matrix. This can be done in one of two ways. The matrix is processed as one long vector where no use is made of the symmetry property. Alternatively, the upper triangular components of the matrix are computed and then copied into the lower elements. In the second case the upper triangle of the matrix is processed row by row as there is no easy way to represent the upper triangle as one long vector.

Since the processing time of a vector on a pipeline machine consists of a start-up time and a vector-processing time the second way leads to inefficiency in the processing of short vectors. Thus the choice between the methods for the construction of symmetric matrices depends on the size of the matrix and the degree of complexity of the operations required to generate the matrix.

Of the four matrix algorithms, none can be efficiently implemented on the Cray-1 in a direct way. For example, take the first algorithm for implementation on the Cray-1. The cascade partial sum method of summing the components of a vector (when the three-dimensional structure is reduced to the two-dimensional structure of the product matrix) is not efficient on the Cray-1 because of the small vector lengths to be processed towards the end of the algorithm completion. However, this algorithm can be modified to the form which can be efficiently implemented on the Cray-1 (Clint *et al.*, 1983).

The second algorithm is more suitable for execution on the Cray-1, particularly for large problems, but then one encounters the general drawback that it is impossible to find an efficient representation of the algorithms in the parallel language suitable for the Cray-1.

12.3   Concluding Remarks

The computation of a matrix product is a process particularly suited for synchronized parallel processing. However, the above analysis of the synchronized matrix product algorithms is conducive to understanding the all-important role that the specific architectural features of parallel structures have in determining the 'best' parallel algorithm. In addition, depending on whether the size of the problem, is less than, equal to, or greater than the number of PEs, different algorithms may be deemed as 'best' on one and the same given parallel architecture.

13

The Parallel FFT Algorithm

The FFT algorithm forms the core of digital signal processing. The convolutions needed to implement digital filters, the correlations needed to implement matched filters, and the Fourier analysis needed for producing spectrograms and medical tomograms can all be performed efficiently by using the FFT algorithm. The extent to which all these operations can be done in real time is often limited by the rate at which the FFT algorithm can be executed.

Since the FFT by virtue of its nature is a highly parallel algorithm, the real-time signal processing is an area where parallel computers can be used particularly advantageously.

13.1   Bergland's Parallel FFT Algorithm

In 1971 Bergland derived algorithms for parallel FFT implementation (Bergland, 1972). The basic parallel FFT algorithm of Bergland computes an N-term discrete Fourier transform in a parallel manner when N can be expressed as the product of two integers r and s which are relatively prime. For simplicity one can always assume that r is odd and s is a power of two. The parallel FFT algorithm then segments the computations of an N-term FFT in an odd number of processes which can be run in parallel.

Let us recall (Chapter 2) that in the DFT problem, one is given N points

$$a_0 \quad a_1 \ldots a_{N-1},$$

and a series of N Fourier coefficients is computed

$$A_0 \quad A_1 \ldots A_{N-1}$$

by means of the formula

$$A_p = \sum_{q=0}^{N-1} w^{pq} a_q \tag{13.1.1}$$

where $w = e^{2\pi i/N}$, $i = \sqrt{-1}$, the unity root.

Assume that N is a product of two factors, r and s, $N = rs$. Then, redefining p and q as follows:

$$p = p_1 + p_2 r, \qquad 0 \leqslant p < N, \qquad 0 \leqslant p_1 < r, \qquad 0 \leqslant p_2 < s$$
$$q = q_2 + q_1 s, \qquad 0 \leqslant q < N, \qquad 0 \leqslant q_2 < s, \qquad 0 \leqslant q_1 < r$$

we derive the following algorithm for computing the Fourier coefficients A_j:
(i) Compute the DFT on r points

$$b_{p_1, q_2} = \sum_{q_1=0}^{r-1} (w^s)^{p_1 q_1} a_{q_1 s + q_2}.$$

(ii) Compute the twiddle factor

$$c_{p_1, q_2} = w^{p_1 q_2} b_{p_1, q_2} . \qquad (13.1.2)$$

(iii) Compute the DFT on s points

$$A_p = \sum_{q_2=0}^{s-1} (w^r)^{p_2 q_2} c_{p_1, q_2}.$$

It is convenient to eliminate the explicit presence of the twiddle factor step, (ii), in the algorithm (13.1.2). One variant of achieving this has been proposed by I.J. Good (Cooley *et al.*, 1967; Good, 1971).

In the Good prime-factor algorithm, q and p of equation (13.1.1) are defined such that

$$q = r q_2 + s q_1, \qquad \text{mod } N, 0 \leqslant q < N,$$
$$q_2 = q r_s (\text{mod } s), \qquad \text{mod } s, \ 0 \leqslant q_2 < s,$$
$$q_1 = q s_r (\text{mod } r), \qquad \text{mod } r, \ 0 \leqslant q_1 < r,$$

and

$$p = s s_r p_1 + r r_s p_2, \qquad \text{mod } N, \ 0 \leqslant p < N,$$
$$p_1 = p (\text{mod } r), \qquad \text{mod } r, \ 0 \leqslant p_1 < r,$$
$$p_2 = p (\text{mod } s), \qquad \text{mod } s, \ 0 \leqslant p_2 < s,$$

where s_r and r_s are solutions of

$$s s_r = q (\text{mod } r), \qquad s_r < r,$$
$$r r_s = q (\text{mod } s), \qquad r_s < s$$

respectively.
With these new definitions for q and p equation (13.1.1) becomes

$$A_p = \sum_{q=0}^{N-1} w^{pq} a_q,$$

$$A_{p_1,p_2} = \sum_{q_2=0}^{s-1} \sum_{q_1=0}^{r-1} w^{pq} a_{q_1,q_2}$$

where

$$w^{pq} = w^{(ss_r p_1 + rr_s p_2)(rq_2 + sq_1)}$$

$$= w^{srs_r p_1 q_2 + srr_s p_2 q_1 + sss_r p_1 q_1 + rrr_s p_2 q_2}$$

$$= (w^{sr})^{(s_r p_1 q_2 + r_s p_2 q_1)} ((w^s)^{ss_r})^{p_1 q_1} ((w^r)^{rr_s})^{p_2 q_2}.$$

Noting that

$$w = e^{2\pi i/N} = e^{2\pi i/rs}$$

and

$$(w)^s = e^{2\pi is/N} = e^{2\pi is/rs} = e^{2\pi i/r} = w_r$$

and similarly

$$(w)^r = w_s,$$

we can write

$$w^{pq} = (w^{sr})^{(s_r p_1 q_2 + r_s p_2 q_1)}(w_r^{ss_r})^{p_1 q_1}(w_s^{rr_s})^{p_2 q_2}.$$

Further, since

$$w^{sr} = (w^{2\pi i/N})^{sr} = 1,$$
$$w_r^{ss_r} = w_r \quad (\text{since } ss_r = q(\text{mod } r)),$$
$$w_s^{rr_s} = w_s \quad (\text{since } rr_s = q(\text{mod } s)),$$

we get

$$w^{pq} = w_r^{p_1 q_1} w_s^{p_2 q_2}.$$

Thus equation (13.1.1) becomes

$$A_{p_1,p_2} = \sum_{q_2=0}^{s-1} \sum_{q_1=0}^{r-1} w_r^{p_1 q_1} w_s^{p_2 q_2} a_{q_1 q_2}$$

$$= \sum_{q_2=0}^{s-1} \left[\sum_{q_1=0}^{r-1} w_r^{p_1 q_1} a_{q_1,q_2} \right] w_s^{p_2 q_2}$$

$$\text{denote by } b_{p_1,q_2}$$

or equivalently:

(i) Compute the DFT on r points

$$b_{p_1,q_2} = \sum_{q_1=0}^{r-1} w_r^{p_1 q_1} a_{q_1,q_2}.$$

(13.1.3)

(ii) Compute the DFT on s points

$$A_{p_1,q_2} = \sum_{q_2=0}^{s-1} w_s^{p_2 q_2} b_{p_1,q_2}.$$

The algorithm (13.1.3) is similar in form to algorithm (13.1.2), but the twiddle factors (ii) or algorithm (13.1.2) have been eliminated. Redefining the q, p, q_2, and q_1 helps to eliminate the explicitly present twiddle factor and instead achieves the same effect by forming cyclic permutations of the subsequences being transformed and by allowing subsequences of the results to be permuted from their normal order. Example 13.1.1 illustrates the process of computing the Fourier coefficients using algorithm (13.1.3).

Example 13.1.1.
Let $N = 12 = r \times s = 3 \times 4$. The given sequence of N points is

$$a_0 \quad a_1 \quad a_2 \quad a_3 \quad a_4 \quad a_5 \quad a_6 \quad a_7 \quad a_8 \quad a_9 \quad a_{10} \quad a_{11}.$$

Using Good's notation we have

$$
\begin{array}{ll}
q = rq_2 + sq_1, & \text{mod } 12, \ 0 \leqslant q < 12, \\
q_2 = qr_s(\text{mod } s), & \text{mod } 4, \ 0 \leqslant q_2 < 4, \\
q_1 = qs_r(\text{mod } r), & \text{mod } 3, \ 0 \leqslant q_1 < 3,
\end{array}
$$

$$
\begin{array}{ll}
p = ss_r p_1 + rr_s p_2, & \text{mod } 12, \ 0 \leqslant p < 12, \\
p_1 = p(\text{mod } r), & \text{mod } 3, \ 0 \leqslant p_1 < 3, \\
p_2 = p(\text{mod } s), & \text{mod } 4, \ 0 \leqslant p_2 < 4,
\end{array}
$$

$$
\begin{array}{llll}
ss_r = q(\text{mod } r), & 4s_r = q(\text{mod } 3), & s_r = 1,4,7,10,\dots \\
rr_s = q(\text{mod } s), & 3r_s = q(\text{mod } 4), & r_s = 3,7,11,\dots.
\end{array}
$$

Thus Step (i) of (13.1.3) gives

a_0		$b_{0,0}$	a_1		$b_{0,1}$
a_4	are used to compute	$b_{1,0}$	a_5	are used to compute	$b_{1,1}$
a_8		$b_{2,0}$	a_9		$b_{2,1}$

a_2		$b_{0,2}$	a_3		$b_{0,3}$
a_6	are used to compute	$b_{1,2}$	a_7	are used to compute	$b_{1,3}$
a_{10}		$b_{2,2}$	a_{11}		$b_{2,3}$

and Step (ii) gives

$$
\begin{array}{llll}
b_{0,0} & & A_0 & \\
b_{0,1} & \text{are used to compute} & A_9 & \\
b_{0,2} & & A_6 & \\
b_{0,3} & & A_3 & \\
\end{array}
\qquad
\begin{array}{llll}
b_{1,0} & & A_4 \\
b_{1,1} & \text{are used to compute} & A_1 \\
b_{1,2} & & A_{10} \\
b_{1,3} & & A_7 \\
\end{array}
$$

$$
\begin{array}{lll}
b_{2,0} & & A_8 \\
b_{2,1} & \text{are used to compute} & A_5 \\
b_{2,2} & & A_2 \\
b_{2,3} & & A_{11} \\
\end{array}
$$

Note that in Example 13.1.1 the three four-point transforms of Step (ii) can be evaluated independently and in parallel via any standard FFT algorithm once the $b_{i,j}$'s terms are available. Note also that the four three-point transforms that compute these $b_{i,j}$ terms can be computed using the formula of Step (i) directly. The samples of the original $\{a_j\}$ series are used in their normal order. Hence in general, the parallel FFT algorithm partitions the computations of an N-term FFT in an odd number of processes which can be run in parallel.

Two types of FFT algorithms are distinguished, called *decimation-in-time* and *decimation-in-frequency*. The decimation-in-time derives its name from the fact that in the process of arranging DFT computations into smaller transformations, the time sequence is decomposed into successively smaller subsequences. In decimation-in-frequency the frequency spectrum is decomposed into smaller subsequences. Subsequently the parallelism of the FFT algorithm can be implemented either with the view of increasing the maximum length of the subsequences that are transformed (decimation-in-time) or with the view of decreasing the execution time (decimation-in-frequency).

In a highly parallel algorithm like the FFT, computation time in various multiprocessor organizations remains the same while the communication overhead due to interprocessor data transfer is extremely important, deciding the actual performance of an algorithm on certain multiprocessor architectures (Lint and Agerwala, 1981).

An implementation of the Bergland parallel FFT algorithm for $N = 12 = r \times s = 3 \times 4$ on the multiprocessor system is illustrated in Fig. 13.1.1. In this system N original terms, $a_0, a_1, \ldots, a_{11}$, are initially stored in the control processor. At the end of the computation they are replaced by its finite DFT.

Each of the r arithmetic processors is capable of performing the s-point transforms of Step (ii) on the data stored in its own memory under the general control of the control processor. The N-term series is presented sequentially by the control processor to the entire ensemble of r arithmetic processors (broadcasting). Each of the sums of Step (i) are accumulated during this time by reserving one memory location in each of the arithmetic processors for this

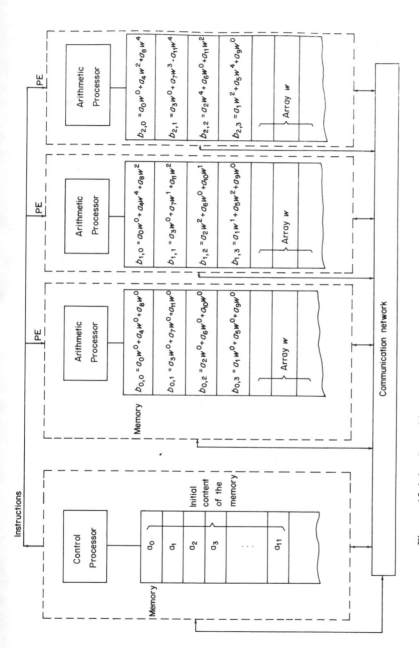

Figure 13.1.1 A multiprocessor system implementation of the FFT algorithm. Intermediate results formed in the PE's own memories. Then $b_{i,j}$'s have to be multiplied by w_4^{00}, w_4^{01}, w_4^{02}, w_4^{03}; w_4^{10}, w_4^{11}, w_4^{12}, w_4^{13}; w_4^{20}, w_4^{21}, w_4^{22}, w_4^{23} respectively, to form the A_p's

purpose. For example, as shown in Fig. 13.1.1, the first processor accumulates $b_{0,0}$, $b_{0,1}$, $b_{0,2}$, and $b_{0,3}$ of the s transforms indicated by Step (i), the second processor accumulates $b_{1,0}$, $b_{1,1}$, $b_{1,2}$, and $b_{1,3}$, and the third processor accumulates $b_{2,0}$, $b_{2,1}$, $b_{2,2}$, and $b_{2,3}$.

Each (arithmetic) processor accumulates these coefficients by simply accepting every sample a_q as it is broadcast to all of the processors, then multiplying it by the appropriate power of w (stored in its w array) and finally adding the resulting product to the correct location in its memory. When all the input terms have been processed, the processors' memories have accumulated the intermediate results as depicted in Fig. 13.1.1. The set of A_p terms can then each be transformed independently and concurrently according to the formula of Step (ii).

After performing the r independent s-point transforms, the Fourier coefficients are transported, in order, to the control processor. This is done by noting the correspondence between the A_p notation and the A_{p_1,p_2} notation where $p_1 = p(\bmod r)$ and $p_2 = p(\bmod s)$. In our example for $N = r \times s = 3 \times 4$, at the end of the computation the results will be stored in the memories of the arithmetic processors, as shown in Fig. 13.1.2.

To send the A_p's to the control processor in order of ascending frequency is done by forming $a(\bmod r)$ counter which identifies the value of p_1 and $a(\bmod s)$ counter which identifies the value of p_2, e.g.

$$\begin{aligned}
k_1 &= 0(\bmod 3) \\
k_2 &= 0(\bmod 4)
\end{aligned} \quad \rightarrow \quad \text{count} = 0 \ \rightarrow A_0,$$

$$\begin{aligned}
k_1 &= 1(\bmod 3) \\
k_2 &= 1(\bmod 4)
\end{aligned} \quad \rightarrow \quad \text{count} = 1 \ \rightarrow A_1,$$

$$\cdots$$
$$\cdots$$
$$\cdots$$

$$\begin{aligned}
k_1 &= 11(\bmod 3) = 2 \\
k_2 &= 11(\bmod 4) = 3
\end{aligned} \quad \rightarrow \quad \text{count} = 11 \rightarrow A_{11}.$$

The Bergland algorithm, as implemented on the SIMD system, is most efficient when r (the number of PEs) is considerably smaller than s. For example, given a number of identical processors that can each efficiently perform $2^{10} = 1024$ point transforms, these processors when interconnected could perform $3 \times 1024 = 3072$ point transforms, $5 \times 1024 = 5120$ point transforms, and $7 \times 1024 = 7168$ point transforms, for only a moderate increase in execution time. However, as r becomes larger and larger, the $N - s$ operations associated with evaluating Step (i) take a larger and larger percentage of the total execution time of the algorithm. Generally, for systems with more than five to seven processors the algorithm becomes time-inefficient.

p_1 (processor number)	0	1	2
p_2 (address in A array)			
0	A_0	A_4	A_8
1	A_9	A_1	A_5
2	A_6	A_{10}	A_2
3	A_3	A_7	A_{11}

Figure 13.1.2 Location of the A_p's where p is a function of p_1 and p_2

13.2   Parallel Radix–2 FFT Computation

The sequential FFT algorithm achieves its lower bound on the number of operations required to compute the DFT on a set of N points, when N is a power of 2, that is, $N = 2^m$. Factorization of the DFT formula in equation (13.1.1) is then carried out using factor 2 as a base or a radix of the algorithm.

Example 13.2.1
Recall the formula

$$A_p = \sum_{q=0}^{N-1} w^{pq} a_q.$$

Now let $N = 8 = 2^3$. Set

$$p = p_1 + 2p_2 + 2^2 p_3 \quad \text{and} \quad q = q_1 + 2q_2 + 2^2 q_3,$$

where p_j's and q_j's take values 0 or 1. We can write

$$A_p = A'(p_1, p_2, p_3)$$

$$= \sum_{q_1=0}^{1} \sum_{q_2=0}^{1} \sum_{q_3=0}^{1} a'(q_1, q_2, q_3) a^{pq_1 + p2q_2 + p2^2 q_3}$$

$$= \sum_{q_1} \sum_{q_2} \sum_{q_3} a'(q_1 q_2 q_3)\, w^{pq_1} w^{p2q_2} w^{p2^2 q_3}$$

$$= \sum_{q_1} \sum_{q_2} \sum_{q_3} a'(q_1, q_2, q_3) w^{pq_1} w_4^{pq_2} w_2^{pq_3}$$

where $w_j = e^{2\pi i/j}$ and $w = e^{2\pi i/N}$.

We now observe that

$$w_2^{pq_3} = w_2^{(p_1+2p_2+2^2p_3)q_3} = w_2^{p_1q_3}w_2^{2p_2q_3}w_2^{2^2p_3q_3} = w_2^{p_1q_3},$$

since

$$w_2^{2p_2q_3} = (e^{2\pi i/2})^{2p_2q_3} = e^{2\pi ip_2q_3} = 1$$

and

$$w_2^{2^2p_3q_3} = (e^{2\pi i/2})^{2^2 p_3q_3} = e^{2\pi i2p_3q_3} = 1,$$

and

$$w_4^{pq_2} = w_4^{(p_1+2p_2+2^2p_3)q_2} = w_4^{p_1q_2}w_4^{2p_2q_2}w_4^{2^2p_3q_2}$$

$$= w_4^{p_1q_2+2p_2q_2},$$

since

$$w_4^{2^2p_3q_2} = (e^{2\pi i/4})^{4p_3q_2} = e^{2\pi ip_3q_2} = 1.$$

Thus

$$A_p = A'(p_1,p_2,p_3)$$

$$(13.2.1)$$

$$= \sum_{q_1} w^{pq_1} \sum_{q_2} w_4^{(p_1+2p_2)q_2} \sum_{q_3} w_2^{p_1q_3}a'(q_1, q_2, q_3)$$

and the following algorithm for computing the Fourier coefficients' A_p's can be deduced:

(i)   Compute
$$c(q_1, q_2, p_1) = \sum_{q_3} w_2^{p_1q_3}a'(q_1, q_2, q_3).$$

(ii)   Compute
$$c(q_1, p_2, p_1) = \sum_{q_2} w_4^{(p_1+2p_2)q_2}c(q_1, q_2, p_1). \qquad (13.2.2)$$

(iii)   Compute
$$c(p_3, p_2, p_1) = \sum_{q_1} w^{pq_1} c(q_1, p_2, p_1).$$

Because of the binary nature of the parameters q_3, q_2, and q_1 each of the sums in algorithm (13.2.2) consists of two terms only, e.g.

$$c(q_1, q_2, p_1) = w_2^0 a'(q_1, q_2, 0) + w_2^{p_1} a'(q_1, q_2, 1).$$

Furthermore, since p_1 takes values 0 or 1 only, the computations of the two terms, $c(q_1, q_2, 0)$ and $c(q_1, q_2, 1)$, can be diagrammed as shown in equation (13.2.3).

$$(13.2.3)$$

Similar diagrams can be drawn for each of the sums in algorithm (13.2.2). The computational diagram (13.2.2) is called the *butterfly* computation of the FFT algorithm (13.2.2), and the algorithm itself is called the radix–2 FFT algorithm.

In algorithm (13.2.2) factorization is based on the parameter q of the data sequence $(a_q, q=0, 1, \ldots, N-1)$; such an algorithm is called a decimation-in-time N point radix–2 FFT algorithm. It is not difficult to see that factorization can be carried out in a similar manner, based on the parameter p of the Fourier coefficients' sequence; the corresponding algorithm is called a decimation-in-frequency N point radix–2 FFT algorithm.

Because of the simplicity of programming the radix–2 algorithm has been extensively used in implementation and hardware considerations in the serial machine environment. The algorithm also lends itself very conveniently to a parallel implementation, not only on a special-purpose FFT computer but also on a general-purpose parallel machine (Bhuyan and Agrawal, 1983).

Using equation (13.2.3) we can represent algorithm (13.2.2) as shown in Fig. 13.2.1.

Note that in the output sequence of the Fourier coefficients, the indices of the elements are obtained by the formula where the binary parameters p_3, p_2, p_1 are used in the reversed order, compared with that of their use in the input data formula, i.e. in the output sequence the element with index $p = p_3 + 2p_2 + 2^2 p_1$ corresponds to the input datum with $p = p_1 + 2p_2 + 2^2 p_3$. This is a well-known phenomenon of the FFT algorithm and it implies that upon completion of the FFT computations, a so-called *unscrambling* (bit reversal) procedure is used to ensure that the results are output in the natural order. In some applications of the FFT (Temperton, 1983), the unscrambling is referred to as a 'self-sorting' procedure, which is incorporated in the FFT algorithm and which ensures that the input to and output from the FFT appear naturally ordered.

It is also seen from Fig. 13.2.1 that in a general decimation-in-time N point radix–2 FFT algorithm, at each stage $N/2$ butterfly computations are carried out and that there are altogether $m = \log_2 N$ such stages. In parallel

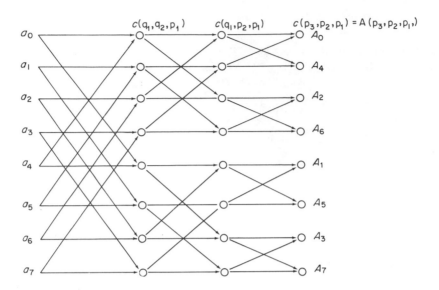

Figure 13.2.1   The FFT algorithm for $N = 2^3$

implementation of the FFT algorithm with p processors, $N/2$ butterfly computations are carried out by each PE per stage. As an example, consider an eight-point FFT which is processed with four PEs, as shown in Fig. 13.2.2. In the example, to keep the diagram simple, the inputs are bit-reversed and the outputs are ordered.

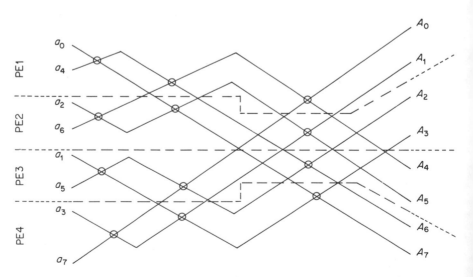

Figure 13.2.2   Eight-point radix-2 FFT computation with four processors

Each PE computes $N/2p$ butterflies per stage, then keeps one datum of each butterfly in its own memory for the next stage computation and sends the other data to some other PE. The process is continued until $m = \log_2 N$ stages are complete.

It is obvious that the performance of this algorithm is strongly influenced by the interprocessor communication type. Bhuyan and Agrawal (1983) have studied implementation of the radix–2 FFT algorithm in multiprocessors with a shared bus and multistage interconnection network (MIN), as proposed by Stone (1971) and Pease (1977), and in MCCs.

To estimate performance of the algorithm with p processors, the speed-up ratio is defined as usual by

$$\sigma = t/T, \tag{13.2.4}$$

where t is the time taken for FFT implementation in a single PE and T is the time taken for FFT implementation in p PEs. Then the 'perfect' speed-up for a computer with p PEs is equal to p.

Also the *cost efficiency* ratio, which is defined as

$$\zeta = \sigma/cp, \tag{13.2.5}$$

where c is a cost factor dependent upon the architecture under consideration, reflects the view that the performance of the algorithm depends heavily on the machine constants. For example, assuming the organization of a shared bus communication network, the cost of a shared bus computer is given as

$$C_s = pC_p + C_c, \tag{13.2.6}$$

where p is the number of PEs, C_p is the cost of PE, and C_c is the cost of the control processor.

It has been shown that for a radix–2 algorithm, the shared bus computer shows close to theoretical performance for $p \leqslant 16$, and a computer with MIN is more cost effective for $p > 16$. When p and N are very large, the shared bus system is completely unsuitable because of high congestion in the single bus. Multiprocessors with MINs are not yet commercially available, but research of this kind indicates their high potential advantage in various types of applications. Both the shuffle exchange network of Stone (1971) and the indirect binary n-cube network of Pease (1977) are ideal for FFT implementation.

13.3 Parallel Computation of the Two-dimensional DFT

There are many signals which are inherently two-dimensional. Also, the FFT techniques may be used to facilitate numerical solution of problems where a multi-dimensional concept is predominant. For example, in the numerical weather prediction models, used for climate research and for day-to-day weather forecasting, the equations on a three-dimensional grid of typically 300 000 points need to be solved (Temperton, 1983). One of the contemporary models of this nature is based on a regular latitude–longitude grid on the sphere. However, it suffers from the drawback that the convergence of the meridians implies progressively shorter grid lengths in the zonal direction as the poles are approached, requiring a prohibitively short time-step for computational stability. To remedy the situation the tendencies of the time-dependent variables are 'Fourier filtered' poleward of a given latitude. This means that the tendencies are expanded in terms of Fourier series which are then truncated at a given wave number, depending on latitude, and finally reconstituted at the grid points. Fourier filtering is implemented via a two-dimensional FFT algorithm.

In such situations the discrete input data are a square matrix

$$
\begin{matrix}
a_{00} & \cdots & a_{0,N-1} \\
\cdot & & \cdot \\
\cdot & & \cdot \\
\cdot & & \cdot \\
a_{N-1,0} & \cdots & a_{N-1,N-1}
\end{matrix}
$$

and a two-dimensional $N \times N$ discrete Fourier transform of $(a_{r,s})$ is given by

$$
A_{u,v} = \sum_{r=0}^{N-1} \sum_{s=0}^{N-1} a_{r,s} w^{ur} w^{vs}, \tag{13.3.1}
$$

$$
w = e^{2\pi i/N}, \qquad i = \sqrt{-1}, \qquad 0 \leqslant u,v \leqslant N-1.
$$

The expression in equation (13.3.1) can very conveniently be decomposed into two one-dimensional DFTs

$$
B_{r,v} = \sum_{s=0}^{N-1} a_{r,s} w^{vs}, \qquad 0 \leqslant r,v \leqslant N-1, \tag{13.3.2}
$$

and

$$
A_{u,v} = \sum_{r=0}^{N-1} B_{r,v} w^{ur}, \qquad 0 \leqslant u,v \leqslant N-1. \tag{13.3.3}
$$

The one-dimensional DFTs (13.3.1) and (13.3.2) are usually implemented using a one-dimensional FFT algorithm.

Consider equation (13.3.2). It can be written

$$
B_{0,v} = \sum_{s=0}^{N-1} a_{0,s} w^{vs}, \qquad 0 \leqslant v \leqslant N-1,
$$

$$B_{1,v} \quad = \quad \sum_{s=0}^{N-1} a_{1,s} w^{vs}, \qquad 0 \leqslant v \leqslant N-1,$$

$$. \\ . \\ .$$

(13.3.4)

$$B_{n-1,v} \quad = \quad \sum_{s=0}^{N-1} a_{N-1,s} w^{vs}, \qquad 0 \leqslant v \leqslant N-1.$$

In equation (13.3.4) each row of the $N \times N$ matrix $(B_{r,v})$ is computed by taking a one-dimensional FFT on a row of the $N \times N$ matrix (a_{rs}).

The set (13.3.3) is treated in a similar way, where for computing each row of matrix $(A_{u,v})$, a column of matrix $(B_{r,v})$ is necessary.

In a parallel implementation of algorithms (13.3.2) and (13.3.3) on p processors, where $p = 2^m$ and $p \leqslant N$, one or more rows of $(a_{r,s})$ are allocated to each PE. After completion of Step (13.3.2), a matrix transpose operation needs to be performed on $(B_{r,v})$ in order to facilitate columnwise distribution of matrix $(B_{r,v})$ between p processors, needed for the processing of Step (13.3.2). This necessary data transfer between the PEs, reduces the speed-up in the performance of the parallel algorithms.

14

Parallel Comparison Sorts

The efficient implementation on parallel systems of the classical comparison problems of merging, sorting, and selection has been very much in attention in recent years (Batcher, 1968; Valiant, 1975; Baudet and Stevenson, 1978; Hirschberg, 1978; Preparata, 1978; Thompson and Kung, 1977; Todd, 1978; Nassimi and Sahni, 1979; Orenstein *et al.*, 1983; Kruskal, 1983).

We already know that the time complexity of comparison sorting of a file of n elements on a single processor is $O(n\log n)$ for both the worst-case and the average-case inputs.

The application of parallel processors is naturally expected to reduce the time requirement. In the extreme case when the number of processors is assumed to be sufficient for every element to be simultaneously compared with every other, all required comparisons at any time can be done in one time unit. This substantially simplifies complexity of any comparison-based algorithm. At the other extreme, the input size becomes very large (infinite) compared with the given number of processors, k. These cases, though of some interest to the theory of computational complexity, are not relevant to practical applications.

For real-life comparison problems, the time complexity of a problem is expressed as a function of the number of processors and of the input data size.

The comparison algorithms are typically developed for the synchronized model of computation, which is a synchronized SIMD system with small local memories of individual processors and a special instruction, the root instruction, which allows a processor to read one location of the memory of an adjacent (in the sense of the interconnection strategy) processor and store its contents in its own memory. Each processor is normally assumed to have the capability of inhibiting the execution of the current instruction by setting an appropriate indicator. The time complexity of parallel comparison algorithms strongly depends on the interconnection pattern between the PEs.

There is very little known yet about asynchronous comparison algorithms.

In terms of the number of parallel processors, the comparison algorithms are distinguished for the three different cases, where the number of processors is assumed to be equal to, greater than, or less than the number of elements in the file to be sorted.

In order to make the parallel computation of the comparison problems worth while in realistic terms, a model of parallel computation for these problems should incorporate the following properties:

168

(a) the number of processors is substantially less than the size of the input file, n;
(b) the number of communication lines attached to each processor is less than n and, preferably, constant;
(c) input to and output from the parallel processor is sequential; this implies that any comparison algorithm must take $O(n)$ time;
(d) memory access conflicts, i.e. simultaneous reading of/writing to the same location by more than one processor, do not occur.

Some major interconnection strategies which have been assumed in the studies of the parallel comparison problems are given under the headings below.

Shared memory model (SMM)

1. Each PE has its own local memory.
2. In addition, a common memory is available to all PEs.
3. To transmit data from PE(i) to PE(j): PE(i) writes the data into the common memory then PE(j) reads it. Transmission time is $O(1)$.
4. Two PEs are not permitted to write into the same word of common memory simultaneously.
5. PEs may or may not be allowed simultaneously to read the same word of common memory. Algorithms that require two or more PEs to read the same common memory word simultaneously will be said to have memory fetch conflicts.

Note that to permit simultaneous memory access by several PEs, SMM requires a large amount of PE to memory connections.

Linearly connected processors

The PEs are logically arranged in a one-dimensional array and each processor can communicate with its two logical neighbours, provided they exist. The simplicity of this interconnection pattern makes it easier to implement parallel algorithms which can be later generalized to a higher-dimensional interconnection network.

Cube-connected computer (CCC)

Assume $p = 2^q$ and let $i_{q-1} \ldots i_0$ be the binary representation of i, $0 \leq i \leq p-1$. Let $i^{(b)}$ be the number whose binary representation is

$$i_{q-1} \ldots i_{b+1} \bar{i}_b i_{b-1} \ldots i_0$$

where $\bar{i}_b$ is the complement of i_b, $0 \leq b < q$. In the cube connected model PE(i) is connected to PE($i^{(b)}$), $0 \leq b < q$. An example of a CCC with eight processors is shown in Fig. 14.1.

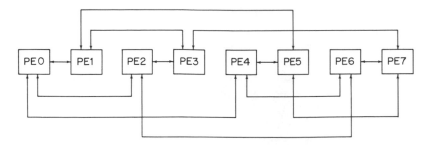

Figure 14.1   A CCC with eight parallel processors

Data can be transmitted from one PE to another only via the interconnection pattern. A CCC requires log p connections per PE.

Perfect shuffle computer (PSC)

Let p, q, i, and $i^{(b)}$ be as in the CCC model. In the PSC, PE(i) is connected to PE($i^{(0)}$), PE($i_{q-2}i_{q-3}\ldots i_0i_{q-1}$) and PE($i_0i_{q-1}i_{q-2}\ldots i_1$). These three connections are normally called exchange, shuffle, and unshuffle (see also Section 9.2). Once again, data can be transmitted from PE to PE only via the connection scheme.

Cube-connected-cycles network

An innovative version of the cube-connected network, a cube-connected-cycles network, has been proposed by Preparata and Vuillemin (1979). In this system each PE is also connected to three other PEs.

Let $n=2^k$ be the number of PEs, where $k=r+2^r$. Each PE has a k-bit address, m, which in turn is expressed as a pair (p, q) of integers represented with $(k-r)$ and r bits respectively, such that $m = p2^r+q$. Each PE(p, q) is connected to PE(p, $(q+1)$mod2^r), PE($p,(q-1)$mod2^r), and PE($p+s2^q$,q), where $s=1-2$BIT$_q(p)$ and BIT$_q(p)$ is the coefficient of 2^q in the binary representation of p.

Thus the PEs are grouped into 2^{k-r} cycles, each cycle consisting of 2^r modules, and the cycles are interconnected as a $(k-r)$ cube. The resulting parallel computer is deemed to require less chip area than one using either the cube-connected interconnection or the perfect shuffle.

14.1   Lower Bounds on Parallel Comparison Problems

Allowing fully arbitrary disjoint comparisons under a parallel synchronized computation model where all interconnection restrictions are relaxed, it is easy to see that the comparison problems have the lower bound of $\log_2 k$ time (see the cascade partial sum method of Section 9.4).

A challenging question is whether this lower bound is achievable under a specific interconnection strategy of a specific parallel model. Only some partial results are known up to date.

To find the maximum of n elements on n processors can be done and requires $\log_2 n$ steps on a network (Knuth, 1973; Batcher, 1968).

Merging two sequences of n elements on n processors again can be done and in $O(\log_2 n)$ time; it will require $O(n \log_2 n)$ comparison (this result is described in Knuth, 1973), and attributed to R. W. Floyd); under the network constraint, this implies a near optimal use of the n processors.

Further, Muller and Preparata (1975) suggested a scheme which sorts n elements with $O(n^2)$ processors in time $O(\log n)$.

More general results on the intrinsic parallelism of the comparison problems of merging, sorting, and finding the maximum are due to Valiant (1975). In particular, an algorithm was developed for merging two sorted sequences of n and m elements, $n \leqslant m$, with $\sqrt{nm}$ processors in $2 \log \log n + O(1)$ comparison steps. Valiant's merging algorithm is described later in this chapter. The general k-processor synchronized computer model solves an n-size comparison problem under the assumption of a linear interconnection network on the PEs.

If the 'speed-up factor' is denoted by P_1/P_k, where P is a given task and P_1 and P_k are measures of the worst-case time complexity of computing the task P, on one and on k processors respectively, then the success of 'parallelization' can be judged by observing how close this speed-up factor is to k. It has been shown that for Valiant's merging algorithm, a speed-up of at least $O(n\sqrt{\log \log n})$ can be achieved.

The results on parallel comparison sorts up to date are summarized in Table 14.1.1. We shall study several basic schemes for parallel sorts. Examples which enhance the exposition of various algorithms are presented in the spirit of Lee (1984).

14.2   Enumeration Sorting (Preparata, 1978)

General idea. Each key is compared with all the others and the number of smaller keys determines the given key's final position.

The algorithm involves three major tasks:

Count acquisition. The set of keys is partitioned into subsets and for each key the number of smaller keys in each subset is determined (in a simple version of the algorithm it is assumed that all keys are distinct).

Rank computation. The final position (rank) of the key in the sorted sequence is determined as the sum of the counts obtained in the count acquisition.

Data rearrangement. Each key is placed in its final position according to its rank.

The simplest version of the algorithm assumes that in the count acquisition phase the set of n keys is partitioned into n subsets with a single key in each

Table 14.1.1 Parallel Comparison Sort Algorithms

Method	Input size	Number of processors	Interconnection network	Time	Space
Enumeration sorts					
Preparata (1978)	n	$n \log n$	Sorting network	$O(\log n) + c \log n$	
Muller and Preparata (1975)	n	$O(n^2)$	Sorting network	$O(\log n)$	
Nassimi and Sahni (1979)	$n \times n$ n	$n \times n$ $n^{1+1/k}$, $1 \leq k \leq \log n$	Mesh Perfect shuffle	$O(k \log n)$	
Ramakrishnan and Browne (1983)	n	n	Linear	$O(n)$	
Odd–even transposition sorts					
Baudet and Stevenson (1978)	n n n	$\log n$ $(\log n/\log \log n)^2$ $2^{\sqrt{\log n}}$	Linear Mesh Perfect shuffle	$O(\log n)$ $O((\log n/\log \log n)^2)$ $2^{\sqrt{\log n}}$	
Thompson and Kung (1977)	n $n \times n$	n $n \times n$	Linear Mesh	$O(n)$ $O(n)$	
Bucket sorts					
Hirschberg (1978)	n numbers in the range of $\{0, \ldots, m-1\}$	n		$O(n)$	$O(mn)$
Orenstein et al. (1983)	n	$O(\log n)$	Linear	$O(n \log n)$	$O(n \log n)$
Mergesorts					
Batcher (1968)	n	$O(n \log^2 n)$	Sorting network	$O(\log^2 n)$	
Valiant (1975)	Two sorted files of size n and m	$\sqrt{nm}$		$2 \log \log n + O(i)$	
Orenstein et al. (1983)	n	$O(\log n)$	Linear	$O(n \log \log n)$	$O(n)$
Todd (1978)	n	$O(\log n)$	Linear	$O(n)$	

subset, then each key is compared with every other key. This requires $O(n^2)$ processors, since the scheme assigns $O(n)$ processors to each key. Example 14.2.1 illustrates the algorithm on a file of 4 elements sorted using 16 processors.

Example 14.2.1

Consider the simplest case of n^2 processors available for sorting a file of n keys. Let $n = 4$. The input file is $(K1, K2, K3, K4) = (4, 2, 1, 3)$.

Count acquisition

Each row of n processors shows an action of comparison of the given key (on the left side of the pair) with every other key of the file:

4:4	4:2	4:1	4:3
2:4	2:2	2:1	2:3
1:4	1:2	1:1	1:3
3:4	3:2	3:1	3:3

If key < otherkey then
    put 0 into the (i, j)th processor
else
    put 1 into the (i, j) processor.

1	1	1	1
0	1	1	0
0	0	1	0
0	1	1	1

At the end of the phase the number of 1's in the *i*th row of the processor-grid indicates how many keys (including *i*th key) numerically precede the *i*th key in the file. This in turn indicates whereabout in the sorted file the *i*th key should be.

Rank computation

Row one (corresponds to the rank of key '4'): $1 + 1 + 1 + 1 = 4.$
Row two (corresponds to the rank of key '2'): $0 + 1 + 1 + 0 = 2.$
Row three (corresponds to the rank of key '1'): $0 + 0 + 1 + 0 = 1.$
Row four (corresponds to the rank of key '3'): $0 + 1 + 1 + 1 = 3.$

Data rearrangement

Place '4' in the fourth position in the output file.
Place '2' in the second position in the output file.
Place '1' in the first position in the output file.
Place '3' in the third position in the output file.

In the count acquisition phase all processors execute in parallel and hence the time taken is constant (independent of n). The rank computation phase uses $O(n^2)$ processors and requires the time bound $O(\log n)$ since it counts in parallel the number of 1's in a set of n binary digits (see the cascade sum method). The final data rearrangement phase simply moves each key to its correct position in the sorted file. By using n processors all keys can be moved in parallel, so the time taken is no more than that required to move a single key and again is independent of n. Thus the time taken to execute the entire algorithm is $O(\log n)$.

Special considerations

The algorithm must ensure that all ranks be distinct, otherwise memory store conflicts would occur. This is particularly important in the general case of some or all keys being identical. In such cases some convention for the ordering of the keys must be adopted. One such convention is to require that the sorting must be stable, i.e. the initial order of identical keys is preserved in the sorted file (as in Knuth, 1973).

If the subsets contain more than one key, count acquisition may be carried out using some merging mechanism to merge the sorted subfiles stored in different processors. Valiant's merging algorithm may be adopted for this purpose (see Section 14.5).

The central idea of the general enumeration algorithm is this:

Let n be the size of the key file and $n = kr$.

The file is first partitioned into subfiles of size r each, which are distributed between k PEs and sorted in order internally by each PE. Let the sorted subfiles be stored in $A_s[0:r-1]$, for $s = 0, 1, \ldots, k-1$.

Now consider two sorted subfiles $A_j[0:r-1]$ and $A_i[0:r-1]$, $r > 1$. Without loss of generality, let $j < i$, and let $B[0:2n-1]$ be the subfile obtained by merging the two sorted subfiles $A_j[0:r-1]$ and $A_i[0:r-1]$ with the ordering convention $A_p[q] \leqslant A_p[q+1]$, $p=i,j$ and $B[q] \leqslant B[q+1]$. Further assume that the rule of stable merging is enforced. Then it is easy to see that

if $B[v] = A_i[w]$ then there are $[v-w]$ entries of $A_j[0:r-1]$ in $B[0:v-1]$ which are no greater than $A_i[w]$, and, similarly, if $B[v] = A_j[y]$ then there are $(v-y)$ entries of $A_i[0:r-1]$ in $B[0:v-1]$ that are strictly less than $A_j[y]$.

Enumeration algorithms were studied under the assumptions that
(a) memory fetch conflicts are permitted;
(b) broadcast capabilities are available (these are needed to accommodate the requirement that a key be simultaneously compared with several other keys);
(c) the overhead of the reassignment of processors to the operation of merging pairs of subsequences is neglected.

The analysis of the algorithm under this model shows that n keys can be sorted in parallel with $n \log n$ processors in time $c \log n + O(\log n)$.

Nassimi and Sahni (1979) have proposed another variant of an enumeration algorithm and studied its implementation for the SMM. Their time-bound results are the same as those for the Preparata algorithm. They have further shown that their algorithm can be implemented on the CCC model and, by simulation, on the PSC; in both cases the implementation assumes $n^{1+1/k}$, $1 \leq k \leq \log n$, processors, and the time requirement of $O(k \log n)$.

The same idea of ranking a key by comparing it against every other key in the file can be used on n linearly connected PEs with a different algorithm, but with the time complexity $O(n)$ (Ramakrishnan and Browne, 1983).

Let **A** and **B** be two arrays, each of size n, and each containing the set of keys to be sorted. Here, **C** is the array that stores the number of keys less than the corresponding key in **B**, i.e. it holds the ranks of the keys in **B**. Initially $C[i] = 0$, for $i = 0, 1, 2, \ldots, n-1$.

The idea is to compare all the keys in **A** and **B** simultaneously (hence n processors are needed). If $A[j] < B[j]$, then $C[j]$ is incremented by one, for $j = 0, 1, \ldots, n-1$. Next, circular left-shift the elements of **A**. Compare again, and so on. The operation is repeated a total number of n times (hence $O(n)$ time complexity), after which each key will have been compared with every other key and **C** will contain the correct ranks of the keys in **B**.

The last operation is the rearrangement of the keys in **B** into **A** according to the ranks.

This brings us to the following implementation of the algorithm.

```
for i := 1 to n do
cobegin j: ⟨0 : n−1⟩
    if A[j] < B[j] then
        C[j] := C[j] + 1
    A[j] := A[j+1] mod n        //left shift one place//
coend

//rearrangement of B into A gives the sorted set//
cobegin j: ⟨0 : n−1⟩
    A[C[j]] := B[j]
coend
```

Example 14.2.2
Let $n = 6$, and

$$B = (6, 3, 4, 5, 2, 1) \quad \text{and} \quad A = (6, 3, 4, 5, 2, 1).$$

$i = 1$:

$$C = (0, 0, 0, 0, 0, 0)$$

left-shift A one place $\qquad A = (3, 4, 5, 2, 1, 6).$

$i = 2$:

    Compare each element of B with the
corresponding element of A
$C = (1, 0, 0, 1, 1, 0)$
left-shift A one place $A = (4, 5, 2, 1, 6, 3)$.

$i = 3$:

    $C = (2, 0, 1, 2, 1, 0)$

$i = 4$: $A = (5, 2, 1, 6, 3, 4)$.

    $C = (3, 1, 2, 2, 1, 0)$

$i = 5$: $A = (2, 1, 6, 3, 4, 5)$.

    $C = (4, 2, 2, 3, 1, 0)$

$i = 6$: $A = (1, 6, 3, 4, 5, 2)$.

    $C = (5, 2, 3, 4, 1, 0)$

                                   $A = (1, 2, 3, 4, 5, 6)$ is the sorted set.

14.3   Odd–even Transposition Sort (Knuth, 1973; Baudet and Stevenson, 1978; Thompson and Kung, 1977)

General idea. Given a file $A[1:t]$ to be sorted, the $A[i]$'s are pairwisely compared, $A[i]:A[i+1]$, and if out of order, exchanged. On pass One through the file, the index i takes on all odd values,

$$i = 1, 3, \ldots, 2\lfloor 2\lfloor (t+1)/2 \rfloor - 1,$$

then on Pass Two, index i takes on all even values,

$$i = 2, 4, \ldots, 2\lfloor t/2 \rfloor.$$

The two passes are used alternately and repetitively until the sorting is complete.

    The algorithm is of minimal extra storage since it requires memory only for the set to be ordered and no space is needed for a second copy of the file. It is shown that the algorithm leads to a completely sorted file after at most t passes (Baudet and Stevenson, 1978). The proof is based on the observation that the distance between the position of a key in the sequence after p passes and its final position is bounded by $t-p$.

    The odd–even transposition is the least efficient of the exchange sorts on a sequential computer. However, it has the characteristics that all comparisons are disjoint. Since there are no common keys between consecutive comparisons, they can be done at the same time very efficiently on a parallel computer.

    In the multiprocessor environment, the simplest version of the algorithm may be used, if the number of keys, k, to be sorted is equal to the number of parallel processors, p_j, $j=1, \ldots, k$. The file is then distributed between k processors, so that each processor contains one key. On Pass One, for $j=1, 3, \ldots, 2\lfloor(k+1)/2\rfloor - 1$, processor p_j compares keys $A[j]$ and $A[j+1]$ and if $A[j] > A[j+1]$ the two keys are exchanged. On Pass Two, the same comparison-exchanges are executed for $j=2, 4, \ldots, 2\lfloor k/2 \rfloor$. Passes One and Two are used alternately and repetitively until the complete file is in order. The algorithm is

```
for i := 1 step 2 to k do
//odd pass//
    cobegin j: ⟨1, 3, ..., 2((k+1) div 2) − 1⟩
        if A[j] > A[j+1] then
            interchange the two keys
    coend
//even pass//
    cobegin j: ⟨2, 4, ..., 2(k div 2)⟩
        if A[j] > A[j+1] then
            interchange the two keys
    coend
enddo
```

In the general case when the number of keys, n, to be sorted is larger than the number of processors, k, the algorithm consists of two typical major phases. In Phase One an unordered file of n keys, $n=kr$, is partitioned into k subfiles of r keys each, $A_j[0:r-1]$, $j=1, 2, \ldots k$. The subfiles are distributed between k processors, p_j, $j=1, 2, \ldots, k$, and each subfile is sorted in order internally by its 'own' processor.

In Phase Two the sorted subfiles are 'merge-splitted' between the processors until the complete file is in order. The idea of the so-called merge-splitting operation was first discussed by Knuth (1973) and then used by Baudet and Stevenson (1978). It consists of two basic steps which are repeated alternatively, and at most k times, until the complete file is in order:

```
for i := 1 step 2 to k do
//odd pass//
    cobegin j: ⟨1, 3, ..., 2((k+1) div 2) − 1⟩
        processor p_j merges the two sorted subfiles A[j] and A[j+1], and then
        assigns to A[j] the first half of the resulting merged file (i.e. the r
        smallest keys) and assigns to A[j+1] the second half
    coend
//even pass//
    cobegin j: ⟨2, 4, ..., 2(k div 2)⟩
        same merge operations are executed for all even j
    coend
enddo
```

Thus the central idea of the generalized algorithm is: given any algorithm using only comparison exchanges for sorting k keys with k processors, there is a corresponding algorithm for sorting rk keys with k processors where every comparison in the first algorithm is replaced by a merge-sorting of two ordered files of r keys in the second.

The algorithm leads to a completely sorted file on n keys in at most k steps (Knuth, 1973, the merging network theorem).

Example 14.3.1

Five steps of parallel odd–even transposition for sorting the partially sorted sequence 55, 63, 78; 13, 68, 81; 27, 93, 99; 17, 31, 43; 38, 47, 84; 40, 54, 67.

$A[1]$: 55 63 78
$A[2]$: 13 68 81
$A[3]$: 27 93 99
$A[4]$: 17 31 43
$A[5]$: 38 47 84
$A[6]$: 40 54 67

$\overset{p1}{\to}$ 13 55 63
68 78 81
$\overset{p3}{\to}$ 17 27 31
43 93 99
$\overset{p5}{\to}$ 38 40 47
54 67 84

Step 1

$\overset{p2}{\to}$ 13 55 63
17 27 31
68 78 81
$\overset{p4}{\to}$ 38 40 43
47 93 99
54 67 84

Step 2

$\overset{p1}{\to}$ 13 17 27
31 55 63
38 40 43
$\overset{p3}{\to}$ 68 78 81
47 54 67
$\overset{p5}{\to}$ 84 93 99

Step 3

13 17 27
31 55 63
38 40 43
68 78 81
47 54 67
84 93 99

$\overset{p2}{\to}$ 13 17 27
31 38 40
43 55 63
47 54 67
$\overset{p4}{\to}$ 68 78 81
84 93 99

Step 4

$\overset{p1}{\to}$ 13 17 27
31 38 40
43 47 54
$\overset{p3}{\to}$ 55 63 67
68 78 81
$\overset{p5}{\to}$ 84 93 99

Step 5

Using the algorithm on the SIMD system under a linear interconnection mechanism, a file of n keys can be sorted with log n processors in time $O(\log n)$, under a mesh-connected scheme, with $(\log n/\log \log n)^2$ processors in time $O((\log n/\log \log n)^2)$ and under a perfect shuffle network, with $2^{\sqrt{\log n}}$ processors in time $O(2^{\sqrt{\log n}})$.

At about the same time Thompson and Kung (1977) have studied a number of parallel sorting algorithms, including the odd–even transposition sort, on a mesh-connected parallel computer. Their model had no provision for wrap-around at the perimeter of the mesh.

In order to simulate different interconnection patterns between the processors on the mesh-connection set-up, different ways of indexing the processors were implemented. These are as shown.

Let the initially loaded input of $n = 16$ keys be

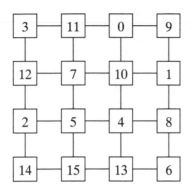

Then:

(i) If the row-major indexing system is used, after sorting is complete, the indexing will be as

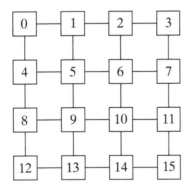

(ii) If the shuffled row-major indexing system is used, the final indexing will be

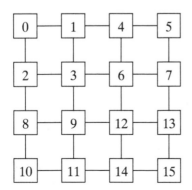

which is obtained by shuffling the binary representation of the row-major index, e.g. the row-major index 5 has the binary representation 0101. After shuffling the bits we have 0011 which is 3. In general *klmnoprs* is shuffled into *kolpmrns*.

(iii) The snake-like row-major indexing is obtained from the row-major indexing by reversing the order of the even rows, i.e.

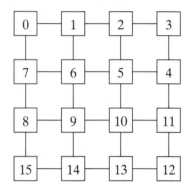

It has been found that the odd–even transposition sort requires an optimal $O(n)$ time on a linearly connected n-processor computer. This linear order connection on the mesh is achieved by mapping two-dimensional arrays with snake-like row-major ordering.

Generally, for other algorithms on the mesh-connected parallel computer studied by Thompson and Kung (1977), it was found that the row-major indexing is poor for merge sorting in general.

14.4 Bucket (Distribution) Sorts (Hirschberg, 1978; Orenstein, *et al.*, 1983)

General idea. The keys are distributed into different 'buckets' by means of some key-examining mechanism or by comparing keys to predetermined values. For the algorithm to be efficient it is very useful to have some information about the structure or the range of the keys.

An interesting mechanism for developing parallel sorting algorithms of this type was suggested by Hirschberg. It demonstrates the three-way trade-off between the processors, time, and space to obtain reasonable lower bounds on the number of processors and on the execution time, with a possible 'penalty' in terms of a greater space requirement by the algorithm.

For convenience the processors are assumed numbered by consecutive integers starting from zero and the algorithm instructions may involve memory references or constants that are a linear function of the bits in the binary representation of the processor number.

Let the n numbers to be sorted $\{c_i, i = 0, \ldots, n-1\}$ are from $\{0, 1, \ldots, m-1\}$ and that duplicate numbers are to be discarded. So $m-1$ is the largest possible number allowed in the file.

In the common memory, m sections of memory, m_j, $j = 0, \ldots, m-1$ are set up, one for each bucket.

Each processor p_i is assigned a number c_i, the ith number being sorted, and accordingly it places the value i in bucket c_i. Then the processor numbers contained in the buckets may be used to activate the corresponding processors to give the sorted file.

The problem with this is that, in general, there may be several values of i with the same number c_i. A memory conflict would result from the simultaneous attempts of several processors to store different values of i into the same bucket.

An answer to this problem is to eliminate duplicate copies of the same number. Processor p_i will be temporarily deactivated if there is another processor p_j where $j < i$ and $c_j = c_i$. This way, for each number to be sorted, only one processor will be active when i is placed in the bucket c_i.

An implementation of the elimination is to allocate n locations (numbered 0 to $n-1$) to each memory section m_j. Within each area m_j, all processors p_i, having $c_i = j$, can leave marks at locations i without fear of memory conflict.

Iteratively, each processor then determines whether or not its 'buddy' is active within the same area. ('Buddy' is defined as in the Buddy System for dynamic memory allocation, Knuth (1973).) For the buddy of lower rank, if it is active then the processor will be deactivated, if not, the processor will continue, shifting its mark to the location of the buddy.

In this way, the processor will shift its mark towards the first location of the area m_j. If there is more than one processor present in the same area, only the processor p_i with the smallest i will reach the first location. And as the other processors are working their way down to the lower-rank buddies, they will eventually be deactivated, at least when they reach the first (numbered 0) location, which is active.

Algorithm parallel-bucketsort

//input: AREA$[j, i] = 0$,   for the processors to leave marks in;
            BUCKET$[j] = -1$,   to hold the processor numbers in the order determined by the actual numbers the processors are holding;

    where
        $0 \leqslant i \leqslant n-1$
        $0 \leqslant j \leqslant m-1$;
    c_i belongs to $\{0, 1, \ldots, m-1\}$ not necessarily distinct;

ouput:
        BUCKET$[j] = (\min i \text{ s.t. } c_i = j)$, $0 \leqslant j \leqslant m-1$;
Let $E_k = 0\ldots010\ldots0$,
        all bits zero except the kth from the right;   //
//Only AREA and BUCKET are in the common memory, all the other

182

variables are local, i.e. there will be one copy of each such variable for each processor.   //

```
cobegin i: ⟨0:n−1⟩
   let i = B_log n ... B_2 B_1
      be the binary representation of i
   x := i              //x is the location that p_i will mark//
   AREA[c_i, x] := 1  //indicates processor p_i's presence//
   active := true      //and is currently active//
   for k := 1 to log n do
   //determines whether or not its buddies are active within the same area//
      buddy := x ⊕ E_k      //address of the buddy, ⊕ is the 'exclusive or'
                              operator, its effect is to complement the kth
                              bit of x//
   count := AREA[c_i, buddy]
   //count ≠ 0 if buddy is active//
   //if buddy is active and if the processor p_i is of higher rank, then p_i will be
   deactivated and the buddy will continue//
      if (x_k = 1 and count ≠ 0) then
      //x_k = the kth bit (from the right) of x//
         active := false
      endif
   //if buddy is not active and is of lower rank, then shift the mark to its
   location//
      if (x_k = 1 and count = 0 and active) then
         AREA[c_i, x] := 0
         x := buddy
         AREA[c_i, x] := 1
      endif
   enddo
   if active then
      BUCKET[c_i] := i
   endif
coend
```

The algorithm requires $O(mn)$ space, the use of n processors, and time $O(\log n)$—there is only one loop which is repeated $\log n$ times. An example will make the algorithm clearer.

Example 14.4.1
Suppose that the range of the numbers is $(0, \ldots, 4)$, i.e. $m = 5$, and an input file contains eight numbers, i.e. $n = 8$. Suppose further that the starting configuration is as follows:

Processor	BUCKET[j]	AREA[j, i]

Processor	BUCKET[j]
p_0 : 2	-1
p_1 : 1	-1
p_2 : 4	-1
p_3 : 2	-1
p_4 : 0	-1
p_5 : 4	
p_6 : 3	
p_7 : 4	

AREA[j, i]

	0	1	2	3	4	5	6	7
0	0	0	0	0	0	0	0	0
1	0	0	0	0	0	0	0	0
2	0	0	0	0	0	0	0	0
3	0	0	0	0	0	0	0	0
4	0	0	0	0	0	0	0	0

A brief dry run of the algorithm

Since it is a parallel algorithm, each of the instructions

$x := i$
AREA[c_i, x] := 1
active := true

will be done simultaneously for all the processors. At the end of the instructions, AREA will look like this:

	0	1	2	3	4	5	6	7
0					1			
1		1						
2	1			1				
3							1	
4			1			1		1

The zeros are omitted for clarity. All the processors are concurrently active. It is not possible to run all the **for** loops simultaneously on paper, so they will be simulated sequentially.

$i = 0 = 000$ (binary representation of i)

	$k = 1$	$k = 2$	$k = 3$
x :	000	000	000
E_k :	001	010	100
	. . .	. . .	. . .
Buddy :	001	010	100

BUCKET[2] := 0

Note that when the processor is already in the first location (numbered 0) it will remain there. All its buddies will be of higher rank and whether active or not, they will be ignored.

$i = 1 = 001$

$$
\begin{array}{cc}
 & k = 1 \\
x : & 001 \\
E_k : & 001 \\
 & \cdots \\
\text{Buddy} : & 000
\end{array}
$$

$x_1 = 1$ and count $= 0$ and active, so processor p_1 shifts its mark to the buddy's location.

The reason for the relation $(x_k = 1)$ in the algorithm is now clearer: because after $x \oplus E_k$, the kth bit is complemented. If $x_k = 0$, it will become 1, and the address of the buddy is of higher rank and so will be ignored. But if $x_k = 1$, the address of the buddy is of lower rank. This is a case of interest as there is a need to decide whether to shift the mark or deactivate the processor.

Now $x = 000$, the rest of the loop will not change anything.
BUCKET[1] := 1

$i = 2 = 010$

$$
\begin{array}{ccc}
 & k = 1 & k = 2 \\
x : & 010 & 010 \\
E_k : & 001 & 010 \\
 & \cdots & \cdots \\
\text{Buddy} : & 011 & 000
\end{array}
$$

Processor p_2 shifts its mark to location 0.
BUCKET[4] := 2

$i = 3 = 011$

$$
\begin{array}{ccc}
 & k = 1 & k = 2 \\
x : & 011 & 010 \\
E_k : & 001 & 010 \\
 & \cdots & \cdots \\
\text{Buddy} : & 010 & 000
\end{array}
$$

After $k = 1$, the processor shifts its mark to the location of its buddy, 010. After $k = 2$, the buddy is of lower rank and is active, so the processor is deactivated and no writing is made into the bucket.

The action of the algorithm should be clear by now. So if there is more than one processor initially active in the same area, only the processor p_i with the smallest i will be able to place its value of i into the bucket.

At the end of the algorithm, the contents of the buckets are:

BUCKET	Sorted file
0 : 4	$p_4 \rightarrow 0$
1 : 1	$p_1 \rightarrow 1$
2 : 0	$p_0 \rightarrow 2$
3 : 6	$p_6 \rightarrow 3$
4 : 2	$p_2 \rightarrow 4$

An extended version of the Bucketsort, which gives the ranking of all input numbers, keeping equal numbers in the same relative order but giving them different ranks, uses n processors and requires $O(mn)$ space and $O(\log n + \log m)$ time.

More recently, Bucketsort algorithms which use only $O(\log n)$ processors and require $O(n)$ or $O(n \log \log n)$ time were proposed by Orenstein et al., (1983). The following algorithm will sort n keys in time $O(n)$ given $O(\log n)$ processors.

Using an order preserving the key-to-address function (Sorenson et al. 1978) the keys are distributed among m buckets, $b_1, b_2, \ldots, b_m$, so that each key in b_i precedes each key in b_j if $i < j$.

Each bucket is sorted by one processor using an $O(n \log n)$ method.

The buckets are concatenated and the sorted file is output.

The algorithm is implemented on a system with linear arrangement of the processors. The processor p_0 stores the input file and receives the result. The processors p_i, $i=1, \ldots, m$, are assigned to buckets b_i respectively. There is a two-way communication line between p_i and p_{i+1}, $i=0, 1, \ldots, m-1$.

During input each record travels from p_0 towards p_m until its destination is reached.

After the records of b_i have been sorted by p_i, the records travel in order back to p_0. If n_i denotes the number of records received by p_i, $i=1, \ldots, m$, then the time requirement of the algorithm ('measured' by the number of parallel comparisons) is

$$T(n) = \max[O(n_i \log n_i)]. \tag{14.4.1}$$

In the worst case of the algorithm, when the keys are all placed in the same bucket, equation (14.4.1) gives

$$T(n) = O(n \log n).$$

Furthermore, each processor used in the algorithm requires $O(n)$ storage space in the worst case, and thus the total memory requirement for the algorithm is $O(n \log n)$.

14.5 Merge Sorts (Batcher, 1968; Valiant, 1975; Todd, 1978; Orenstein *et al.*, 1983)

Merging is commonly used for sorting. It is specially useful in a parallel computer environment. We have seen that, given n keys and n processors, there are some good algorithms to sort the keys, e.g. odd–even transposition or Bucketsort. However, n will be very large in practice, e.g. 10^8. Processors are costly to build, so it would not be economical to have so many processors. A realistic model should require a sublinear number of processors, say $k = \log n$. A simple method is to put n/k numbers into each of the k processors, then parallel merge of the sorted subfiles will give the final sorted file. Hence the general idea.

One assumes two or more sorted subfiles of keys. Then several pairs (or multiples) of keys, one from each sorted subfile, are compared simultaneously and, if out of order, exchanged. The output is a sorted file of keys.

In 1968 Batcher proposed two parallel merge-sorting algorithms, which use a sorting network of 'comparators' and are based on the principle of repetitive merging. His work was one of the earliest major results in the area of parallel sorting.

A sorting network is a hardware unit consisting of a number of comparison units, many of which can operate in parallel. A network is designed to sort a specific number of inputs, and the measure of a network is the number of comparisons required to sort the given number, n, of keys. A great deal of theoretical attention has been given to the problem of minimizing the number of comparisons required to sort some specific n. The concept of a sorting network is explained on an example of six numbers in Fig. 14.5.1. The pattern of comparisons in a sorting network can be adopted by a general-purpose parallel computer. The essence of such a method of sorting is the development of comparison sequences which are disjoint so that a number of processors can operate as independent comparison units. The overhead of the method is the development of the proper sequences.

The Two-way Odd–even Mergesort

Batcher's parallel Mergesort operates in a number of passes. Preliminary passes arrange keys to form an ordered file of odd-numbered keys and an ordered file of even-numbered keys, so that during the last pass no key will be further than one position away from its proper place.

Keys of the odd file and even file are neighbours at the last pass and the files are merged by comparing each key with its neighbour. All the comparisons of the last step are disjoint and can be done in parallel.

Let the two sorted subfiles of keys $(A_1, A_2, \ldots, A_m)$ and $(B_1, B_2, \ldots, B_n)$ be merged into one sorted file $(C_1, C_2, \ldots, C_m)$. The algorithm is

Phase One.

if m = n = 1 **then**
a single comparison will sort the numbers
else if $mn > 1$ **then**
    merge the odd sequences $(A_1, \ldots, A_{2\lceil m/2 \rceil - 1})$ and
    $(B_1, \ldots, B_{2\lceil m/2 \rceil - 1})$, obtaining $(F_1, \ldots, F_{\lceil m/2 \rceil + \lceil n/2 \rceil})$;
    concurrently with merging of the even sequences
    $(A_2, \ldots, A_{\lfloor m/2 \rfloor})$ and $(B_2, \ldots, B_{\lfloor m/2 \rfloor})$ obtaining $(E_1,$
    $\ldots, E_{\lfloor m/2 \rfloor + \lfloor n/2 \rfloor})$
    endif
endif

Phase Two.

Interleave the two merged sequences to obtain
$(F_1, E_1, F_2, E_2, \ldots, F_{\lfloor m/2 \rfloor + \lfloor n/2 \rfloor},$
$E_{\lfloor m/2 \rfloor + \lfloor n/2 \rfloor}, F', F'')$,
where
$F' = F_{\lfloor m/2 \rfloor + \lfloor n/2 \rfloor + 1}$ does not exist if both m and n are
    even,
$F'' = F_{\lfloor m/2 \rfloor + \lfloor n/2 \rfloor + 2}$ does not exist if both m and n are
    odd.

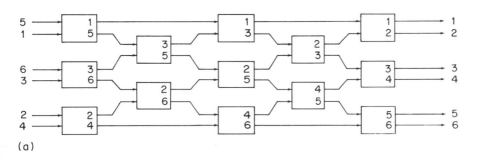

(a)

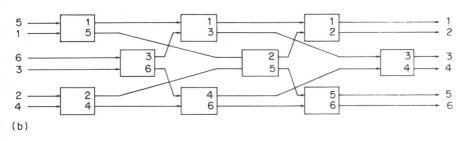

(b)

Figure 14.5.1   A sorting network for sorting the sequence 5, 1, 6, 3, 2, 4. A network consists of a number of comparators of which each has two inputs (represented by incoming arrows) and two outputs (outgoing arrows); the higher input is the smaller of the two inputs and the lower input is the larger. (a) A sorting network with five parallel comparisons; (b) a sorting network with three parallel comparisons

Phase Three.   Apply a single parallel comparison-interchange step to:
$$E_1 : F_2,$$
$$E_2 : F_3,$$
$$E_3 : F_4,$$
$$\dots$$
$$E_{\lfloor m/2 \rfloor + \lfloor n/2 \rfloor} : F'$$
will produce the sorted result.

The lowest output of the odd merge is left alone and becomes the lowest number of the final file.

The ith output of the even merge is compared with the $(i + 1)$st output of the odd merge to form the $2i$th and $(2i+1)$st numbers of the final file ($i = 1, 2, \dots, \lfloor m/2 \rfloor + \lfloor n/2 \rfloor$).

If the output remains in the odd or even merge it is left alone and becomes the highest number in the final file.

Algorithm for Linearly Connected Processors

For the purpose of this and the following sections, only two instruction types are needed: the routing instruction on two data elements in each processor and the comparison instruction, which is a conditional interchange on the values of two registers in each processor. In fact, both types of such instructions are needed to allow either register to receive the minimum.

Define:

T_r = time required for one unit-distance routing step, i.e. moving one item from a processor to one of its neighbours;

T_c = time required for one comparison step.

Concurrent data movement is allowed, as long as it is all in the same direction. It means that a comparison-interchange step between two items in adjacent processors can be done in $2T_r + T_c$ (route left, route right) time. A mixture of horizontal and vertical comparison interchanges will require at least $4T_r + T_c$ time.

An example will be used to illustrate how odd–even Mergesort can be done on a linearly connected multiprocessor.

Example 14.5.1
Let the two sorted subsequences $(1, 3, 4, 6)$ and $(0, 2, 5, 7)$ be initially loaded in the first and second halves of the eight linearly connected processors:

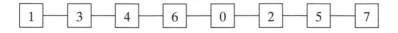

(i) Unshuffle: move the odd-indexed elements to the left and the evens to the right:

(ii) Merge the 'odd sequences' and the 'even sequences':

(iii) Perfect shuffle: interleave the two merged susequences:

(iv) Comparison-interchange (an interchange is indicated by ↔):

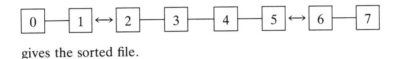

gives the sorted file.

Note that the perfect shuffle can be achieved by using the triangular interchange pattern:

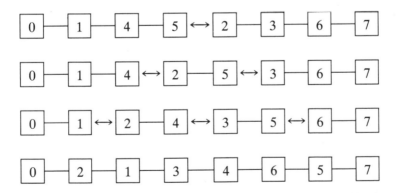

Similarly, an inverted triangular interchange pattern will do the unshuffling. Therefore, both the perfect shuffle and the unshuffling can be done in $n/2-1$ interchanges, i.e. $n-2$ routing steps, when performed on a row of length n.

190

Algorithm for Mesh-connected Processors

A mesh-connected array of PEs is assumed with no 'wrap-around' connections, i.e. end processors have two or three rather than four neighbours:

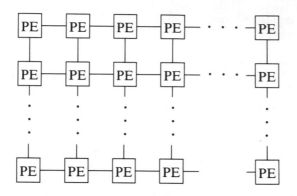

Let $M(j, k)$ denote an algorithm of merging two j by $k/2$ sorted adjacent subarrays to form a sorted j by k array, where j, k are powers of 2, $k > 1$, and all the arrays are arranged in the snake-like row-major ordering.

Example 14.5.2
    First consider the case when $k = 2$. Given the two sorted columns of length $j > 1$, say, $j = 4$:

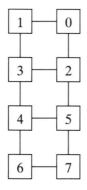

Algorithm $M(j, 2)$ is as follows:

T1. Move all odds to the left-hand column and all evens to the right:

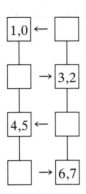

time = $2T_r$ because route left and route right are to be performed separately.

T2. Use the 'odd–even transposition sort' to sort each column:

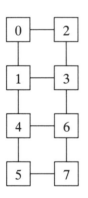

time = $j(2T_r + T_c)$ because j comparison-interchanges are needed to sort j items.

T3. Interchange on even rows:

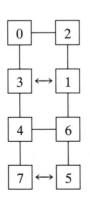

time = $2T_r$. This is the interleave step in snake-like order.

T4. One step of comparison-exchange (every 'even' with the next 'odd'):

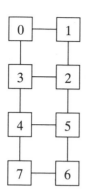

time $= 2T_r + T_c$. Now the array is sorted in a snake-like row-major order. Note the total time needed is $T(j, 2) = (2j + 6)T_r + (j + 1)T_c$.

Example 14.5.3
The 2-way odd–even merge. For $k > 2$, say, $k = 4$:

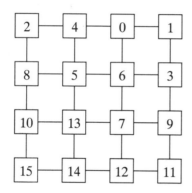

$M(j, k)$ is defined recursively:

M1. Single interchange step on even rows if $j > 2$, so that columns contain either all evens or all odds. If $j = 2$, do nothing; the columns are already segregated:

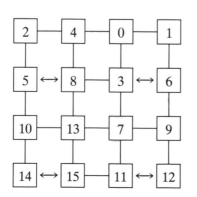

time $= 2T_r$. For an array with two columns and in snake-like ordering, the indexing is
odd → even
even ← odd,
odd → even,
even ← odd.
So, to achieve all odds and all evens in each column, the even rows need to be interchanged.

M2. Unshuffle each row:

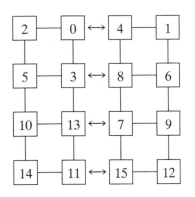

time = $(k-2)T_r$. This brings all the odd-indexed numbers to the left half and all the even-indexed numbers to the right half of the array.

M3. Merge by calling $M(j, k/2)$ on each half:

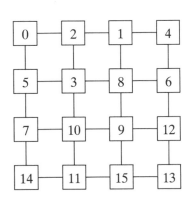

time = $T(j, k/2)$. Note that the left and right halves of the array are both sorted in a snake-like ordering.

M4. Shuffle each row:

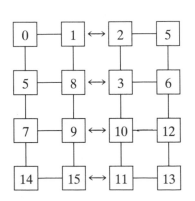

time = $(k-2)T_r$. Note, this is interleaving the odds and evens in row-major order. So the even rows are in the wrong order for snake-like ordering.

194

M5. Interchange on even rows:

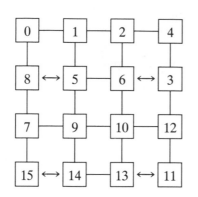

time $= 2T_r$. This extra step is needed to ensure that shuffling is in snake-like row-major ordering.

M6. Comparison interchange of adjacent elements (every 'even' with the next 'odd'):

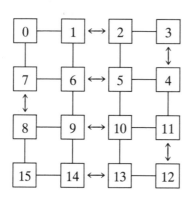

time $= 4T_r + T_c$. Now the whole file is sorted.

Let $T(j, k)$ be the time needed by $M(j, k)$. Then, adding all the times as given above, we get:

$$T(j, k) = T(j, k/2) + (2k + 4)T_r + T_c,$$

and, further, recursively

$$T(j, k) = 2k \sum_{i=0}^{m-1} (1/2^i)T_r + (4T_r + T_c)m + T(j, k/2^m),$$

where $k/2^m = 2$, i.e. $m = \log k - 1$, and by the previous result,

$$T(j, 2) = (2j + 6)T_r + (j + 1)T_c.$$

Thus

$$T(j, k) = 4k(1 - 2/2^{\log k})T_r$$
$$+ (4 \log k - 4 + 2j + 6)T_r + (\log k + j)T_c$$
$$= (4k + 4 \log k + 2j - 6)T_r + (\log k + j)T_c$$

or

$$T(j, k) \leqslant (4k + 4 \log k + 2j)T_r + (\log k + j)T_c. \qquad (14.5.1)$$

An $n \times n$ sort may be composed of $M(j, k)$ by sorting all columns in $O(n)$ routes and comparisons, then using $M(n, 2), M(n, 4), M(n, 8), \ldots, M(n, n)$ for a total of $O(n \log n)$ routes and comparisons.

This rather poor performance is due to two drawbacks in the algorithm:
(i) the recursive subproblems, $M(n, n/2)$ and $M(n, n/4)$, dimensions: they all are $O(n)$ in complexity:
(ii) the method is very 'local' in the sense that no comparisons are made between elements initially in different halves of the array until the last possible moment, when each half has already been sorted.

The first drawback can be remedied by designing many 'upwards' merges to complement the 'sideways' merge just described. This idea is used below.

The 2s-way Merge

Let $M'(j, k, s)$ be an algorithm for merging $2s$ arrays of size j/s by $k/2$ in a $j \times k$ region of processors, where j, k, s are powers of $2, s \geqslant 1$, and the arrays are in snake-like row-major ordering. The algorithm $M'(j, k, s)$ is almost the same as $M'(j, k)$ except that M1 and M6 are replaced by

M'1. Single interchange step on even rows of $j > s$, so that columns contain either all evens or all odds. If $j = s$, do nothing.

M'6. Perform the first $2s-1$ parallel comparison interchange steps of the odd–even transposition sort on the snake. At most

$$\text{time} = s(4T_r + T_c) + (s-1)(2T_r + T_c) = (6s-2)T_r + (2s-1)T_c.$$
↑   Note: When starting comparisons with the 'evens', there will be horizontal as well as vertical comparisons, hence the more expensive $4T_r$ term.

Similarly, for $M'(j, 2, s), j > s$, the Step T4 is replaced by M'6 and will take $(2s-1)(2T_r + T_c)$ time.

It can be seen that the 2-way merge is a special case of the 2s-way merge.

Example 14.5.4

Let $j = k = 8$ and $s = 2$, that is, we wish to merge four arrays of size 4×4 in an 8×8 region of processors. Let the initial four arrays of numbers be:

```
17  18  19  10 | 49  50  51  52
24  23  22  21 | 56  55  54  53
25  26  27  28 | 57  58  59  60
32  31  30  29 | 64  63  62  61
---------------+---------------
 1   2   3   4 | 33  34  35  36
 8   7   6   5 | 40  39  38  37
 9  10  11  12 | 41  42  43  44
16  15  14  13 | 48  47  46  45
```

M′1. Single interchange step on even rows:

```
17    18    19    20  | 49    50    51    42
23↔24 21↔22           | 55↔56 53↔54
25    26    27    28  | 57    58    59    60
31↔32 29↔30           | 63↔64 61↔62
---------------------+----------------------
 1     2     3     4  | 33    34    35    36
 7↔ 8  5↔ 6           | 39↔40 37↔38
 9    10    11    12  | 41    42    43    44
15↔16 13↔14           | 47↔48 45↔46
```

M′2. Unshuffle each row (all the odd columns are on the left):

```
17  19  49  51 | 18  20  50  52
23  21  55  53 | 24  22  56  54
25  27  57  59 | 26  28  58  60
31  29  63  61 | 32  30  64  62
---------------+---------------
 1   3  33  35 |  2   4  34  36
 7   5  39  37 |  8   6  40  38
 9  11  41  43 | 10  12  42  44
15  13  47  45 | 16  14  48  46
```

M′3. Merge by calling M′(8, 4, 2) on each half:

M′1. Interchange on even rows:

```
17    19    49    51  | 18    20    50    52
23↔21 53↔55           | 22↔24 54↔56
25    27    57    59  | 26    28    58    60
29↔31 61↔63           | 30↔32 62↔64
---------------------+----------------------
 1     3    33    35  |  2     4    34    36
 5↔ 7 37↔39           |  6↔ 8 38↔40
 9    11    41    43  | 10    12    42    44
13↔15 45↔47           | 14↔16 46↔48
```

M'2. Unshuffle each row (each half is done separately):

```
17  49  19  51  │  18  50  20  52
23  53  21  55  │  22  54  24  56
25  57  27  59  │  26  58  28  60
29  61  31  63  │  30  62  32  64
 1  33   3  35  │   2  34   4  36
 5  37   7  39  │   6  39   8  40
 9  41  11  43  │  10  42  12  44
13  45  15  47  │  14  46  16  48
```

M'3. Merge by calling M'(8, 2, 2) on each half (of the previous halves):

T1. Move all odds to the left column and all the evens to the right:

```
17, 49          │  19, 51         │  18, 50         │  20, 52
        23, 53  │          21, 55 │          22, 54 │          24, 56
25, 57          │  27, 59         │  26, 58         │  28, 60
        29, 61  │          31, 63 │          30, 62 │          32, 64

 1, 33          │   3, 35         │   2, 34         │   4, 36
         5, 37  │           7, 39 │           6, 38 │           8, 40
 9, 41          │  11, 43         │  10, 42         │  12, 44
        13, 45  │          15, 47 │          14, 46 │          16, 48
```

T2. Sort each column:

```
 1   5  │   3   7  │   2   6  │   4   8
 9  13  │  11  15  │  10  14  │  12  16
17  23  │  19  21  │  18  22  │  20  24
25  29  │  27  31  │  26  30  │  28  32
33  37  │  35  39  │  34  38  │  36  40
41  45  │  43  47  │  42  46  │  44  48
49  53  │  51  55  │  50  54  │  52  56
57  61  │  59  63  │  58  62  │  60  64
```

T3. Interchange on even rows:
T4. Comparison-interchange:

```
 1   5  │   3   7  │   2   6  │   4   8
13   9  │  15  11  │  14  10  │  16  12
17  23  │  19  21  │  18  22  │  20  24
29  25  │  31  27  │  30  26  │  32  28
33  37  │  35  39  │  34  38  │  36  40
45  41  │  47  43  │  46  42  │  48  44
49  53  │  51  55  │  50  54  │  52  56
61  57  │  63  59  │  62  58  │  64  60
```

$M'(8, 2, 2)$ is now finished. But still under the recursive call of $M'(8, 4, 2)$ in the 'first' M'3:

M'4. Shuffle each row (on each halves):

1	3↔ 5	7	2	4↔ 6	8
13	15↔ 9	11	14	16↔10	12
17	19↔23	21	18	20↔22	24
29	31↔25	27	30	32↔26	28
33	35↔37	39	34	36↔38	40
45	47↔41	43	46	48↔42	44
49	51↔53	55	50	52↔54	56
61	63↔57	59	62	64↔58	60

M'5. Interchange on even rows:

M'6. Comparison interchange:

1	3	5	7	2	4	6	8
15	13	11	9	16	14	12	10
17	19	21	23	18	20	22	24
31	29	27	25	32	30	28	26
33	35	37	39	34	36	38	40
47	45	43	41	48	46	44	42
49	51	53	55	50	52	54	56
63	61	59	57	64	62	60	58

Note that all the odds and the evens have been separately sorted.

Now the 'original' M'3 is done.

M'4. Shuffle each row:

1	2	3	4	5	6	7	8
15	16	13	14	11	12	9	10
17	18	19	20	21	22	23	24
31	32	29	30	27	28	25	26
33	34	35	36	37	38	39	40
47	48	45	46	43	44	41	42
49	50	51	52	53	54	55	56
63	64	61	62	59	60	57	58

M'5. Interchange on even rows:

1	2	3	4	5	6	7	8
16↔15		14↔13		12↔11		10↔ 9	
17	18	19	20	21	22	23	24
32↔31		30↔29		28↔27		26↔25	
33	34	35	36	37	38	39	40
48↔47		46↔45		44↔43		42↔41	
49	50	51	52	53	54	55	56
64↔63		62↔61		60↔59		58↔57	

M'6. Comparison-interchange:
   For this particular example, after M'5 the whole file is already sorted, so M'6 is redundant.

The s^2-way Merge

The s^2-way merge $M''(j, k, s)$ is a further generalization of the 2-way merge $M(j, k)$. Input to $M''(j, k, s)$ is $s \times s$-sorted j/s by k/s arrays in a $j \times k$ region of processors, where j, k, s are powers of 2 and $s > 1$. $M''(j, k, s)$ is exactly the same as $M'(j, k, s)$ except that in Step M'6 the expression $(2s-1)$ is replaced by (s^2-1).

   Here, $M''(j, s, s), j \geqslant s$, is a special case analogous to $M(j, 2)$ and can be performed in the following way:

N1. $(\log s/2)$ 2-way merges of $M(j/s, 2), M(j/s, 4), \ldots, M(j/s, s/2)$.
N2. A single $2s$-way merge $M'(j, s, s)$.

It can be shown that if the snake-like row-major indexing is used, the s^2-way merge algorithm may be developed to sort $n \times n$ elements in time

$$[6n + O(n^{2/3}\log n)]T_r + [n + O(n^{2/3}\log n)]T_c,$$

which is optimal within a factor of 7 for all n.

Valiant's Mergesort

A faster Mergesort algorithm has been proposed by Valiant (1975). The method uses $k = \sqrt{mn}$ processors to merge two sorted files of length m and n, where $1 < m \leqslant n$, in time $2 \log \log n + \text{const.}$

   The essence of the algorithm is that given $\sqrt{mn}$ processors, in two time intervals, the problem of merging two files of length m and n can be reduced to one of merging a number of pairs of files, where in each pair the shorter of two files has length less than $\sqrt{m}$. The pairs of files are so created and the $\sqrt{mn}$ processors are so assigned to them at the next stage, that for each pair there will be enough processors allocated to apply the same algorithm repeatedly until the whole file is sorted.

The algorithm is as follows. Given the sorted files

$$X = (x_1, x_2, \ldots, x_m),$$
$$Y = (y_1, y_2, \ldots, y_n),$$

(a) Mark the elements of X that are subscripted by $i\sqrt{m}$ and those of Y subscripted by $i\sqrt{n}$, for $i = 1, 2, \ldots,$ and label them as x'_i and y'_i respectively. Call the subfiles between successive marked elements and after (or before) the last (or first) marked element in each file 'segments'.

(b) Compare each x'_i with each y'_i to decide for x'_i the segment of Y into which it needs to be inserted. This requires no more than $\sqrt{mn}$ comparisons and can be done in unit time (one time unit = one parallel comparison).

(c) Compare each x'_i with every element of the segment of Y that has been found for it to identify where each of the x'_i belongs in Y. This requires at most $\sqrt{m}(\sqrt{n} - 1) < \sqrt{mn}$ comparisons and can also be done in unit time.

(d) Merge the disjoint pairs of subfiles $(X_1, Y_1), (X_2, Y_2), \ldots$ where each X_i is a segment of X and therefore of length $X_i \leqslant \sqrt{m}$ and Y_i is the subfile of Y between the two insertion points of x'_{i-1} and x'_i.

It is shown that this step requires at most $\sqrt{mn}$ processors, and, in general, in the algorithm the repetitive process of successively splitting a pair of files into a set of pairs of subfiles can continue with the given number of processors.

Mergesort of Sorenson *et al.*

This Mergesort is a variant of Sorenson's Bucketsort which eliminates the possibility of all keys occurring in the same bucket by assigning, in an arbitrary fashion, n/m or m/n keys to each PE, where m can be reasonably set to $\log n$. Each PE (bucket) is then sorted by one processor as in the bucketsort variant, and the sorted subfiles are merged rather than concatenated as concatenation will no longer be plausible since the keys are initially distributed in an arbitrary fashion.

The merge is done by a single processor with m input lines, one from each of the buckets (PEs). Assuming that the merge of the sorted subfiles is done by one of the efficient algorithms of time complexity $O(n \log m)$, the algorithm would take $O(n \log \log n)$ comparisons to complete the sort.

The number of connection lines, m, can be reduced if the merge is carried out by a *binary tree of processors*, where each component PE has two input lines. Assuming that the subfile of size n_i, $i=1, 2, \ldots, m$, is sorted by a $O(n_i \log n_i)$-worst-case sort, the time complexity of this Mergesort is $O(n)$ since the time of the merge by a tree of PEs is given as

$$\sum_{i=0}^{\log n - 1} cn/2^i = O(n).$$

15

Parallel Graph Search and Traversal

On serial computers the algorithms for solving problems involving tree and graph structures are particularly time consuming. As we have seen, in general, sequential graph-solving algorithms have exponential time complexity and form a large group of the apparently intractable problems.

An improvement on the time bounds required to solve such problems is conceivable by developing parallel algorithms, especially as the tree and graph problems are inherently parallel. For example, a parallel traversal method of a general graph based on sequential depth-first search could, in theory, cover the graph in time linearly proportional to the traversal of the diameter of the graph (the diameter of a graph is the longest of the shortest paths between all pairs of vertices) while a sequential algorithm can do no better than visit each vertex in turn. Generally, there are many graph procedures which can logically be separated into independent parts; it is then reasonable to execute these parts simultaneously rather than sequentially.

15.1 Distributed Computer Systems (DCS) Graphs

Trees and graphs are widely used in the formulation of many diverse problems; however, we shall consider their application to one particular area only, the area of parallel computers, so that a graph in our setting may represent a distributed computing network (Chang, 1982; Chang and Roberts, 1979). A vertex in a tree or in a graph is an autonomous processor and an edge between two vertices, a channel of interaction/message transmission between the two vertices. A basic distributed computer system (DCS) model is an interconnection of processing elements each having certain capabilities, communicating with other elements through a network, and working on a set of related or unrelated tasks. In particular, in a loosely coupled system the communication time between processors is an important system parameter.

In this set-up, a distributed program is a program consisting of several modules or tasks which are free to reside on any processor in a distributed system. Each processor is characterized by its performance factors, e.g. execution and communication speed, and by its reliability factors, e.g. failure probability, checkpoint time, and restart time.

A general optimization problem of a DCS is an optimal utilization of the system, that is, assigning the processors so as to minimize the execution time of the entire program. The assignment of tasks to processors to maximize performance is called the *load balancing* in a distributed system.

One important DCS characteristic is localized information processing by a subset of the processing elements. A piece of information may be processed repeatedly within a local subsystem before it changes locality. For example, a seat assignment file in an airline reservation system is very heavily used first at the airport where passengers are preparing to board. Then, as the plane flies to another airport, the use locality changes.

To maximize the efficiency of the system, the data file should be distributed so that it is accessible at the locality of first use and should be allowed to migrate as the need changes. The distribution of DCS information is known as the *distribution design problem*. Solutions to the distribution design problems depend on the operations performed on the information. In the simplest form, an operation is a data file access from a specified origin. In a database distributed on a DCS, an operation or query may originate in a program located anywhere in the system. It may access multiple data files assembled at a single vertex before it is processed, or the query and the intermediate results created in each step may be sent sequentially through the files. A combination of the two strategies is also possible.

The general solution to the optimal utilization of the system problem is very hard to find. However, a number of algorithms for the scheduling, distribution design, and load balancing in distributed systems have been proposed (Chou and Abraham, 1982; Wah, 1984).

Algorithms which solve these different graph problems are often based on the fundamental algorithms of graph/tree traversal and search (Chang, 1982), matrix multiplication (Dekel, *et al.*, 1979), or topological sorting (Er, 1983). For example, in the DCS environment, graph solutions may determine activities of control functions in a distributed system, such as:

(a) assigning an identity to a new processor;
(b) finding the ordering of all processor identities in the system;
(c) finding the configuration of the system;
(d) providing a mutual exclusion mechanism which permits only sequential access to a critical set of resources;
(e) finding the clusters of vertices (subgraphs) which are interconnected by single links;
(f) broadcasting a message quickly to all vertices.

We shall consider some of these fundamental algorithms.

15.2 Parallel Graph Algorithms

Fully parallel algorithms on graphs exhibit some basic problems, pertinent to graph problems:

If several edges lead to one vertex and the parallel traversal of edges starting from some initial vertex should arrive at that vertex simultaneously, how is this to be handled?

Does the message from each of the edges get passed on and if not, what is to be done with the ones which are aborted?

How does information get back to the starting vertex in a coordinated fashion?

Different algorithms resolve these problems by enforcing a set of specific rules, which deal with the conflicts in situations arising.

Parallel Graph Traversal

One basic idea of the parallel graph traversal is a straightforward parallel generalization of the sequential depth-first approach: start at one initiator vertex (if a *single source graph*) or simultaneously at all initiator vertices (if a *multisource graph*), and propagate the signals in parallel along all out-edges of the vertex until all sink vertices are reached.

Many of the proposed parallel graph algorithms are founded on a parallel depth-first method. These include a parallel depth-first traversal of a DCS graph of k processors sharing common memory (Eckstein and Alton, 1977; Arjomandi and Corneil, 1979) and the graph algorithms for a distributed system network of finite state machines which assume a synchronized system with simultaneous transitions based on sensing the states of all neighbours at each step (Rosenstiehl *et al.*, 1972).

Chang (1982) has developed several graph algorithms, based on the parallel depth-first method, for a model of a fully decentralized, asynchronous, and with no common memory multiprocessor system. The model assumes that:

(a) each processor is capable of supporting multiple local processes, so that while some application tasks are being executed, it is also capable of sending and receiving messages and of initiating control algorithms;

(b) each processor has message–passing capability, which sends in parallel to the immediate neighbours and is capable of positive acknowledgement of messages received, has enough memory to store all incoming messages;

(c) there is no overtaking between messages and no fixed speed of transmission;

(d) a vertex in the system does not know the extent or membership of the entire graph.

In such a system decentralized algorithms, executed on all processors, must operate asynchronously through message-passing to coordinate the system as a whole.

We shall outline a single-source connected graph traversal algorithm of Chang, where the process starts at some initiator vertex and is continued in parallel, along all out-edges of this vertex, each message carrying the identity of the starting vertex.

The algorithm, called the pure traversal, can operate on a connected single-source undirected graph, or on a strongly connected digraph, i.e. a digraph in which every vertex can be reached from every other vertex.

In the algorithm, traversal of the graph means passing messages from one vertex to another; for any particular vertex v at which the execution of the algorithm starts, the messages originating from v form a family, sharing the identity of v in common; there are two phases in the traversal of a graph, a forward phase and an echo phase; explorers accomplish the forward traversal of a graph and echoes the echo phase; explorers and echoes carry information with them about those parts of the graph which they have traversed; a vertex which synchronizes echoes processes this information and sends the result with the echo from that vertex; the starting vertex finally receives all its echoes from its out-edges, and after processing this information, obtains the result of the algorithm.

The algorithm consists of four phases.

Phase 1. Initiator s sends explorers in parallel on all its out-edges. (For a vertex-receiver the out-edges are all edges except the edge at which the first explorer arrived.)

Phase 2. Upon the arrival of the first explorer at the vertex, the edge is marked as first. Explorers are sent in parallel on the out-edges.

Phase 3. Upon the arrival of each subsequent (after the first) explorer, the echo is sent along the edge on which the explorer arrived.

If more than one explorer arrive at an unvisited vertex simultaneously, only one is chosen as the first explorer to the vertex and its edge of arrival as the first edge of the vertex; the other explorers are then considered as subsequent explorers to a visited vertex. If necessary, an echo-merge arbiter mechanism at each vertex gives some arbitrary sequential ordering to the messages which arrived simultaneously.

Phase 4. Upon the arrival of an echo to the vertex, the edge is marked as having received an echo.

> **if**   all echoes for the vertex have arrived
>   **and** the vertex is not the initiator
> **then**   an echo is sent along the first edge
> **else**   **if** the vertex is the initiator
>   **then** we are finished
>   **endif**
> **endif**

Example 15.2.1 illustrates the algorithm on a DCS graph with 16 processors.

Example 15.2.1

Given is a distributed graph with 16 vertices (processors), fully decentralized. For simplicity, we assume that a message takes one unit of time to travel along any one edge of the graph.

Stage 1.

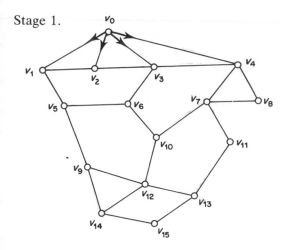

Initiator vertex v_0 sends four copies of a message (four explorers) in parallel to its neighbour vertices v_1, v_2, v_3, and v_4 along its out-edges. v_1, ... v_{15} are unvisited.

Stage 2.

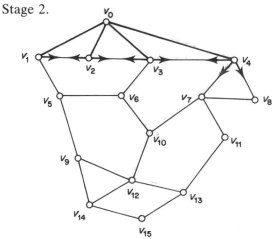

v_1, ..., v_4 each send explorers in parallel.
v_5, ..., v_{15} are unvisited.

Stage 3.

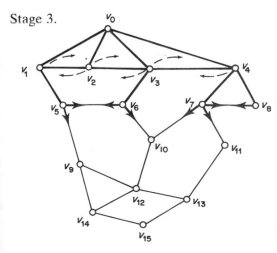

v_1 receives the explorer from v_2, acquires the status of 'visited' and sends an echo-message to v_2; similarly, v_2 sends an echo-message to v_1; v_3 and v_4 are engaged in a likewise action.

v_5, ..., v_8 send an explorer each in parallel on their out-edges.

Stage 4.

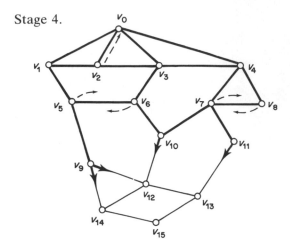

v_2 has received all its expected echoes and so sends an echo along its first edge to v_0.

v_1, v_3, and v_4 are still waiting to receive echoes along some of their out-edges.

$v_9, \ldots, v_{11}$ send explorers in parallel on their out-edges.

Explorers from v_6 and v_7 arrive simultaneously at v_{10} and an explorer from v_6 is chosen as the first (with the help of some selection mechanism).

Stage 5.

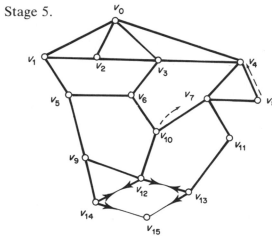

v_8 has received all its echoes and sends an echo on its first edge to v_4.

v_1, v_3, ..., v_7, v_9, v_{10} and v_{11} are still waiting to receive echoes on some of their out-edges.

v_{12}, v_{13}, and v_{14} send explorers in parallel.

v_{15} is unvisited.

Stage 6.

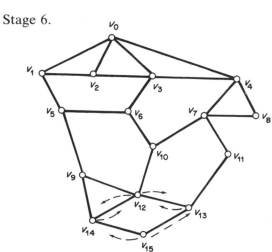

v_2 and v_8 are not in action.

v_{15} is visited; there are no out-edges; it sends an echo on its first edge to v_{13}.

Remaining vertices are waiting to receive echoes.

The travel of echoes along the first edges is shown by a broken line.

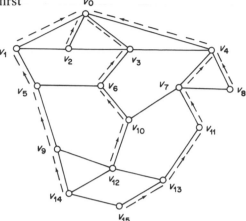

Efficiency of the Traversal Algorithm

Efficiency of the algorithm in terms of the elapsed communication time depends on the relative speed of explorer and echo messages and on the processor time versus communications time. In general, assuming particular conditions on the speed of explorers and echoes, an execution of the pure traversal algorithm may cause different sequences of arrival of explorers and different edges marked the first edges. If, for the purpose of basic analysis of the algorithm, we assume that (i) the processor time is very small compared to communications time, (ii) the messages take approximately the same time, one unit, to traverse any one edge, and (iii) explorers and echoes have the same speeds, then the communications time in the execution of a pure traversal algorithm from an initiator s, is bounded from above by twice the weighted diameter of the graph, where the metric of weight is message travel time. This result follows immediately from the definition of diameter of a graph and the parallel activity of the algorithm.

Other useful characteristics for assessing the efficiency of a graph traversal algorithm are the total number of message passes in the system and the amount of storage required at each vertex. For the pure traversal algorithm the number of message passes in an undirected graph is bounded by $4e$, where e is the number of edges in the graph, since each edge can have at most two explorers, one in each direction, and two corresponding echoes. For a digraph the number of message passes is clearly $2e$, since there are no symmetrical pairs of explorers which travel on directed edges.

Finally the storage requirements of the pure traversal algorithm at each vertex are $O(n)$ bits, where n is the number of vertices in the graph. This result is derived by observing that a vertex w has at most n edges, each of which needs one bit to mark the arrival of its echo (note that the vertex notes the arrival of all echoes before sending its own echo along the first edge); then to mark the first edge of w requires $\log n$ bits and $\log n$ bits are needed to maintain the name of the vertex. In addition, each message carries the basic identity of the initiator and a type, which is ($\log n + 1$) bits. Finally, one bit is needed to mask a vertex as visited. Hence, in total the requirement is ($n+3 \log n + 2$) bits or $O(n)$ bits.

Other parallel algorithms on graphs

A number of decentralized parallel graph algorithms can be derived by modifying the pure traversal algorithm. Some of them are as follows:

1. A *simple single-source algorithm*, where the task is to obtain the identities of all the processors in the system in sorted form; a particular vertex initiates the algorithm which executes in parallel and the required answer is finally obtained by the initiator vertex.
2. A *simple multiple-source algorithm*, where not one but several vertices may initiate an activity which has a common global end. All vertices that initiate the algorithm within a certain functional time bound participate in the algorithm, and the algorithm produces a distributed ordering by identity (or priority) of the participants.
3. A *non-trivial single-source algorithm for finding the biconnected components* of an undirected and connected graph. The answer will be held by a member of each biconnected component and hence is in distributed form.

For a connected undirected graph G, its *biconnected components* are those subsets of vertices and edges which share a common cycle. There may be edges in G which belong to no cycles and these are considered to be a biconnected component of two members. A distinction is made for the so-called *articulation points* of G (Aho, *et al.*, 1974).

A vertex a is called an *articulation point* of G if there exist vertices v and w such that v, w, and a are distinct and every path between v and w contains the vertex a. Biconnected components contain no internal articulation points of G, but themselves must be connected through articulation points of the graph otherwise they would form a single biconnected component. So, though a biconnected component does not contain any internal articulation points, one or more of its vertices may be articulation points of the whole graph G. It is not difficult to see that articulation points are critical in the reliable functioning of a network in that a vertex failure of an articulation point produces a disjoint graph. It is therefore important for a network to discover periodically its articulation points and the members of its biconnected components.

The reader is invited to work out his/her own versions of these algorithms and then, for comparison, consult Chang (1982).

Properties of the DCS graphs

Many graph properties are important for DCS problems, and the algorithms which estimate parametric values of these properties are frequently used in optimal DCS graph algorithms. We have earlier made use of the graph diameter; it is helpful when deriving performance characteristics of DCS graphs. Other useful graph parameters are as follows:

(i) *All pairs shortest paths.* The all pairs shortest path matrix $\mathbf{A}$ is an $n \times n$ matrix such that $A(i, j)$ is the length of a shortest path from i to j.

If $A^k(i, j)$ is the length of a shortest path from i to j going through at most k intermediate vertices then $A(i, j) = A^n(i, j)$.

(ii) *The median and the median length.* Let $d(i, j)$ be the length of a shortest path from i to j. Let $h(j)$ be the weight of vertex j. Vertex v is a weighted median of the graph if and only if

$$\sum_{j=1}^{n} h(j)d(v, j) \leqslant \sum_{j=1}^{n} h(j)d(k, j), \qquad 1 \leqslant k \leqslant n.$$

When $h(j) = 1$, $1 \leqslant j \leqslant n$, the vertex is simply a median of the graph.

Here, $\sum_{j=1}^{n} h(j)d(v, j)$ is called the weighted median length of the graph.

(iii) *The spanning tree of the graph.* The minimum (depth) spanning tree (MST) of a (weighted) graph is defined as that subset of edges of the graph which connects all vertices (with minimum total edge weight).

Other spanning trees of interest are the shortest path spanning tree, the breadth-first spanning tree, and a shortest median spanning tree.

Efficient parallel algorithms for computing different properties of graphs have been proposed by Dekel *et al.* (1979, 1981). The algorithms are based on parallel matrix multiplication algorithms.

15.3 Parallel Topological Sort (Er, 1983)

Definition Given a directed acyclic graph $G = (V, E)$, the topological sort is a computation of a linear order on the vertices, subject to the constraints of partial ordering embedded in the graph.

A number of sequential algorithms for topological sort are known (Kahn, 1962; Knuth, 1973; Knuth and Szwarcfiter, 1974; Reingold *et al.*, 1977; Varol and Rotem, 1981).

Er (1983) has proposed a parallel approach to this problem, based on a simple combinatorial concept of the propagation of vertex values from all source vertices to all sink vertices in a given digraph. Once a vertex is visisted, all links leading from the vertex are traversed down simultaneously. Inductively, all vertices in the digraph will be visited if all source vertices are visited. If the digraph is acyclic, the parallel traversal will terminate as the number of vertices in a digraph is finite.

The algorithm consists of two parts: in part one the pairs of vertices with partial orderings between them are input, and a digraph representing these orderings is set up; in part two topological sorting is computed using a parallel approach. The algorithm is as follows:

repeat
  Read in a pair of partial ordering (v, w)
  if $v < w$
  then place a directed link from n_w to n_v
  else place a directed link from n_v to n_w
  endif
until no more pairs
Initialize all vertex values to zero.
Find all the source vertices of the digraph.
if no source vertex exists //the digraph is cyclic//
then exit
else
  Visit all the source vertices and change their values to 1.
  repeat
  Follow down the directed links from all the vertices n_p just visited and
  visit all of their successor vertices n_s, simultaneously.
  //Update the vertex value of each successor vertex n_s//
  if (the value of n_s) $\leq$ (the value of n_p)
  then (the value of n_s) $= 1 +$ (the value of n_p)
  endif
  until (the computation has converged
    or the value of a vertex is assigned a value larger than the total
    number of vertices in the digraph).
  //The latter signifies that the digraph is cyclic and should be reported
  accordingly.//
    List all the vertices in ascending order of vertex values.
  endif

The algorithm can be implemented on a SIMD machine with common and local memory facilities. To minimize the common memory contention in running the algorithm, each vertex of the digraph is assigned to a common memory module. Then values of all vertices can be updated in parallel.

In order to avoid a memory access conflict the execution of the algorithm can be synchronized. So, within the second **repeat-until** loop, all processors complete one stage iteration before starting the next iteration. Two possibilities may arise: (a) at the end of the current iteration a common successor vertex is reached by two or more processors chaining down the same distance from the source vertices; and (b) in a later iteration a common successor is reached by the processor chaining down the longer path.

In case (a) there can be interference and only one of the processors must increment the vertex value. In case (b) there is no interference. It is shown that provided all processors are synchronized and each vertex value is stored in a memory module, the parallel topological sort is free of contention when implemented on a SIMD machine.

It is also shown that the time complexity of the algorithm is of the order of the longest distance between a source vertex and a sink vertex in an acyclic digraph.

Another parallel topological sort algorithm was proposed by Dekel *et al.* (1979, 1981). The algorithm is developed for a SIMD machine, with n^3 synchronized PEs and with either CCC or PSC interconnection strategies. It does not, however, assume the use of common memory for transporting the intermediate data.

The algorithm requires $O(\log_2 n)$ time.

Concluding remarks

The 1980s are a time for exploration in parallel computing. The developments in numerical analysis, computational complexity and software engineering are following closely the developments in parallel hardware. The issues are many. Are reasonable models for parallel computation being defined? How to provide a bridge from sequential to parallel programming, so that the maximum benefits can be gained from the parallelism intrinsic to particular applications? How to analyze sequential algorithms to identify potentially parallel tasks, what tools are needed for this analysis? What kinds of parallelism are available in the applications algorithms, so that the hardware can be designed to preserve and provide an efficient implementation for that parallelism? Getting the data to the right place at the right time seems to be a major problem in a parallel programming; what new conceptual solutions are available to resolve this problem?

Researchers throughout the world are attacking these problems with great enthusiasm, and a significant and steady progress in the area is recorded continuously. The purpose of this book will be accomplished if the reader finds the presented material an informative introduction to the new fascinating field of parallel processing.

Bibliography and References

Aho, A. V., Hopcroft, J. E., and Ullman, J. D., 1974, *The Design and Analysis of Computer Algorithms*, Addison-Wesley, Reading, Mass.

Aho, A. V., Hopcroft, J. E., and Ullman, J. D., 1983, *Data Structures and Algorithms*, Addison-Wesley, Reading, Mass.

Andrews, G. R., and Schneider, F. B., 1983, Concepts and notations for concurrent programming, *ACM Comput. Surveys,* **15** (1), 3–44.

Ansiello, G., and Marchetti-Spaccamela, M. P., 1980, Toward a unified approach for the classification of NP-complete optimization problems, *Theoretical Comp. Sci.,* **12**, 83–96.

Arjomandi, E. R., and Corneil, D. G., 1978, Parallel computations in graph theory, *SIAM J. Comput.,* **7**, 230–237.

Azar, A., and Gageot, Y., 1983, Vectorization of explicit multidimensional finite difference and finite elements schemes, *EDF.-Bull. de la Direction des Etudes et des Recherches,* Série C, **1**, 23–30.

Barnes, J., 1965, An algorithm for solving nonlinear equations based on the secant method, *Comp. J.,* **8**, 66–72.

Batcher, K. E., 1968, Sorting networks and their applications, *Proc. of the AFIPS Spring Joint Conf.,* **32**, 307–314.

Baudet, G. M., 1978, Asynchronous iterative methods for multiprocessors, *J ACM,* **25**, 226–244.

Baudet, G., and Stevenson, D., 1978, Optimal sorting algorithms for parallel computers, *IEEE Trans. Comput.,* **C–27** (1), 84–87.

Bender, E. A., 1974, Asymptotic methods in enumeration, *SIAM Rev.,* **16** (4), 485–515.

Bentley, J. L., and Shamos, M. I., 1976, Divide-and-conquer in multi-dimensional space, *Proc. of the Eighth Annual ACM Symp. on Theory of Computing,* pp. 220–230.

Bergland, G. D., 1972, A parallel implementation of the fast Fourier transform algorithm, *IEEE Trans. Comput.,* **C–21**, 366–370.

Berlekamp, E. R., 1970, Factoring polynomials over large finite fields, *Math. Comp.,* **24**, 713–735.

Bhuyan, L. N., and Agrawal, D. P., 1983, Performance analysis of FFT algorithms on multiprocessor systems, *IEEE Trans. Softw. Eng.,* **SE–9** (4), 512–521.

Bitton, D., DeWitt, D. J., Hsaio, D. K., and Menon, J., 1984, A taxonomy of parallel sorting, *ACM Computing Surveys,* **16** (3), 287–318.

Blum, M., 1967, A machine independent theory of the complexity of recursive functions, *J ACM,* **14**, 322–336.

Blum, M., Floyd, R. W., Pratt, V., Rivest, R. L., and Tarjan, R. E., 1973, Time bounds for selection, *J. Comput. System Sci.,* **7**, 448–461.

Book, R. V., 1975, Formal language theory and theoretical computer science, *2nd GI Conf. on Automata Theory and Formal Languages, Lecture Notes in Comp. Sc. 33,* Springer-Verlag, pp. 1–15.

Borodin, A., 1969, Complexity classes of recursive functions and the existence of complexity gaps, *Conf. Rec. ACM Symp. on Theory of Computing,* pp. 67–68.

Borodin, A., 1973a, Computational complexity: theory and practice, in *Currents in the Theory of Computing,* A. V. Aho, ed., Prentice-Hall, Englewood Cliffs, N.J., pp. 35–89.

Borodin, A., 1973b, On the number of arithmetics to compute certain functions—circa May 1973, in *Complexity of Sequential and Parallel Numerical Algorithms,* J. F. Traub, ed., Academic Press, New York, pp. 149–180.

Bracewell, R. N., and Riddle, A. C., 1967, Inversion of fan-beam scans in radio astronomy, *Astrophys. L., 150,* 427–434.

Brent, R. P., 1973, Some efficient algorithms for solving systems of nonlinear equations, *SIAM J. on Num. Analysis, 10,* 327–344.

Brocard, O., Bonnet, C., Vigneron, Y., Lejal, T., and Bousquet, J., 1983, A vectorized finite element method for the computation of transonic tridimensional potential flows, *EDF.-Bull. de la Direction des Etudes et des Recherches,* Série C, **1,** 45–50.

Brockett, R. W., and Dobkin, D., 1978, On the optimal evaluation of a set of bilinear forms, *Linear Algebra and its Applications, 19,* 207–235.

Brooks, R. A., and Di Chiro, G., 1976, Principles of computer assisted tomography (CAT) in radiographic and radioisotopic imaging, *Phys. Med. Biol., 21* (5), 689–732.

Brown, M. R., 1977, The analysis of a practical and nearly optimal priority queue, Ph.D. thesis, STAN–CS–77–600, Comp. Sci. Dept., Stanford Univ., Stanford, Calif.

Cantor, G., 1874, see, for example, E. V. Huntington, *The Continuum and Other Types of Serial Order.* (With an Introduction to Cantor's Transfinite Numbers.) 2nd ed. Dover books on mathematics, New York, 1955.

Chang, E. J. H., 1982, Echo algorithms: Depth-first parallel operations on general graphs, *IEEE Trans. Softw. Eng.,* **SE–8** (4), 391–401.

Chang, E., and Roberts, R., 1979, An improved algorithm for decentralised extrema-finding in circular configurations of process, *Comm. ACM, 22* (5), 281–283.

Chazan, D., and Miranker, W.L., 1969, Chaotic relaxation, in linear algebra and applications, **2,** 207–217.

Chou, T. C. K., and Abraham, J. A., 1982, Load balancing in distributed systems, *IEEE Trans. Softw. Eng.,* **SE–8** (4), 401–412.

Church, A., 1932, A set of postulates for the foundations of logic, *Ann. Math., 33,* 346–366; **34,** 839–864.

Church, A., 1936, An unsolvable problem of elementary number theory, *Am. J. Math., 58,* 345–63.

Churchhouse, R. F., 1983, The past, present and future of computers and computer science, *Bull. IMA, 19,* 130–134.

Clint, M., Perrott, R., Holt, C., and Stewart, A., 1983, The influence of hardware and software considerations on the design of synchronous parallel algorithms, *Software—Practice and Experience, 13,* 961–974.

Cobham, A., 1964, The intrinsic computational difficulty of functions, in *Proc. 1964 Internat. Congress for Logic Methodology and Philosophy of Sci.,* Y. Bar-Hillel, ed., North-Holland, Amsterdam, pp. 24–30.

Cohen, H., and Lenstra, H. W. Jr., 1982, Primality testing and Jacobi sums, *Report 82–18,* University of Amsterdam, Dept. of Math.

Cook, S. A., 1971, The complexity of theorem-proving procedures, *Proc. 3rd Annual ACM Symp. Theory of Computing,* Shaker Hts. Ohio, pp. 151–158.

Cook, S. A., 1973, A hierarchy for nondeterministic time complexity, *J. Comput. System Sci., 7,* 343–353.

214

Cook, S. A., 1978, An overview of computational complexity, *Comm. ACM,* **26,** 400–408.

Cook, S. A., and Reckhow, R. A., 1973, Time bounded random access machines, *J. Comput. System Sci.,* **7,** 354–375.

Cooley, J. W., Lewis, P. A., and Welch, P. D., 1967, The fast Fourier transform algorithm and its applications, *IBM Res. Paper RC–1743;* IBM Thomas J. Watson Research Center, Yorktown Heights, New York.

also The fast Fourier transform and its application to time series analysis, in *Statistical Methods for Digital Computers,* K. Enslein, A. Ralston, and H. S. Wilf, eds., Wiley, New York, pp. 377–423.

Cooley, J. W., and Tukey, J. W., An algorithm for the machine calculation of complex Fourier series, in *Math. Comput.,* **19** (90), pp. 297–301.

Coppersmith, D., and Winograd, S., 1982, On the asymptotic complexity of matrix multiplication, *SIAM J. Comput.,* **11,** 472–492.

Cutland, N. L., 1980, *Computability. An Introduction to Recursive Function Theory,* Cambridge University Press.

Dantzig, G. B., 1963, *Linear Programming and Extensions,* Princeton University Press, Princeton, N.J.

Davis, M., and Hersh, R., 1973, *Hilbert's Tenth Problem,* Scientific American, p. 84.

Davis, M., Matijasevic, Y., and Robinson, J., 1976, Hilbert's tenth problem. Diophantine equations: positive aspects of a negative solution, mathematical developments arizing from Hilbert problems, *American Mathematical Society Monthly,* Providence, R.I., pp. 323–378.

Dekel, E., Nassimi, D., and Sahni, S., 1979, Parallel matrix and graph algorithms, *Proc. 17th Annual Allerton Conf. on Comm., Control and Computing,* 10–12 Oct., pp. 27–36; see also *SIAM J. Comput.,* **10** (4), 1981, 657–675.

Dijkstra, E. W., 1959, A note on two problems in connection with graphs, *Numer. Math.,* **1,** 269–271.

Dijkstra, E. W., 1968, Co-operating sequential processes, in Programming Languages, F. Genuys, ed., Academic Press, London, pp. 43–112.

Donath, W. E., 1979, Placement and average interconnection lengths of computer logic, *IEEE Trans. Circuits Syst.,* **CAS–26,** 272–276.

Donath, W. E., 1980, Complexity theory and design automation, *IBM Research Report, RC 8203* (No. 35656), IBM Thomas J. Watson Research Center, Yorktown Heights, New York.

Dongarra, J. J., 1983, Redesigning linear algebra algorithms. *E.D.F.-Bull. de la Direction des Etudes et des Recherches,* Série C—Math. Informatique, **1,** 51–60.

Dubois, M., and Briggs, F. A., 1982, Performance of synchronized iterative processes in multiprocessor systems, *IEEE Trans. Softw. Eng.,* **SE–8,** (4), 419–431.

Eckstein, D. M., and Alton, D. A., 1977, Parallel graph processing using depth-first search, *Proc. Conf. Theoret. Comput. Sci.,* Univ. Waterloo, Waterloo, Ont., Canada, pp. 73–81.

Edmonds, J., 1965, Paths, trees and flowers, *Canad. J. Math.,* **17,** 449–467.

Er, M. C., 1983, A parallel computation approach to topological sorting, *Comp. J.,* **26** (4), 293–295.

Eriksen, O., and Staunstrup, J., 1983, Concurrent algorithms for root searching, *Acto. Inf.,* **18** (4), 361–376.

Evans, D. J., 1983, New algorithms in linear algebra, *EDF.-Bull. de la Direction des Etudes et des Recherches,* Série C–Math. Informatique, **1,** 61–70.

Fisher, M. J., and Rabin, M. O., 1974, Super-exponential complexity of Pressburger arithmetic, in *Complexity of Computation, SIAM–AMS Proc. 7,* R. Karp, ed., pp. 27–42.

Floyd, R., 1967, Nondeterministic algorithms, *J ACM,* **14,** 636–644.

Floyd, R. W., 1972, Permuting information in idealized two-level storage, in

Complexity of Computer Computations, R. Miller and J. Thatcher, eds., Plenum Press, New York.

Flynn, M. J., 1966, Very high speed computing systems, *Proc. IEEE,* **14**, 1901–1909.

Garey, M. R., and Johnson, D. S., 1979, *Computers and intractability. A guide to the theory of NP-completeness,* W. H. Freeman and Co., San Francisco.

Gödel, K., 1931, Uber formal unentscheidbare satze der principia mathematica und verwandter system I. *Monatschefte Math. Phys.,* **38**, 173–198. (English translation in M. Davis, ed., *The Undecidable, Raven,* New York.)

Good, I. J., 1971, The relationship between two fast Fourier transforms, *IEEE Trans. Comput.,* **C–20**, 310–317.

Grzegorczyk, A., 1953, Some classes of recursive functions, *Rozprawy Matematyczne* IV, Warszawa, pp. 1–46.

Guibas, L., McCreight, E., Plass, M., and Roberts, J., 1977, A new representation for linear lists, *Proc. Ninth Annual ACM Symp. on Theory of Computing,* ACM, New York, pp. 49–60.

Hartmanis, J., and Hopcroft, J. E., 1971, An overview of the theory of computational complexity, *J ACM,* **18**, 444–475.

Hartmanis, J., and Simon, J., 1975, On the structure of feasible computations, in GI–4, Jahestagung, Berlin, 9–12 Oktober 1974, Herausgegeben im Auftrag der GI von D. Siefkes, IX. *Lecture Notes in Comp. Sci. 26,* Springer-Verlag, pp. 3–51.

Herman, G. T., 1979, Principles of reconstruction algorithms, *Med. Image Processing Group Tech. Report No. MIPG27,* State Univ. of New York at Buffalo.

Hirschberg, D. S., 1978, Fast parallel sorting algorithms, *Comm. ACM,* **21**, 657–661.

Hockney, R. W., and Jesshope, C. R., 1981, *Parallel Computers,* Adam Hilger, Bristol.

Hopcroft, J. E., and Kerr, L. R., 1971, On minimizing the number of multiplications necessary for matrix multiplication, *SIAM J. Appl. Maths.,* **20** (1), 30–36.

Hopcroft, J., and Tarjan, R., 1973a, Dividing a graph into triconnected components, *SIAM J. Comput.,* **2**, 135–158.

Hopcroft, J., and Tarjan, R., 1973b, Algorithm 447: efficient algorithms for graph manipulation, *Comm. ACM,* **16**, 372–378.

Horowitz, E., and Sahni, S., 1978, *Fundamentals of Computer Algorithms,* Computer Science Press, Inc., Potomac, Maryland.

Hayafil, L., and Kung, H. T., 1975, Bounds on the speed-up of parallel evaluation of recurrences, *Proc. Second USA–Japan Computer Conf.,* pp. 178–182.

Johnson, D. B., 1977, Efficient algorithms for shortest paths in sparse networks, *J ACM,* **24**, 1–13.

Kahn, A. B., 1962, Topological sorting of large networks, *CACM,* **5**, 558–582.

Kaltofen, E. A., 1982a, A polynomial-time reduction from multivariate to bivariate integer polynomial factorization, *Proc. 14th ACM Symp. in Theory Comp.,* San Francisco, Calif., pp. 261–266.

Kaltofen, E. A., 1982b, A polynomial-time reduction from bivariate to univariate integral polynomial factorization, *Proc. 23rd IEEE Symp. on Found. of Comp. Sci.,* IEEE Comp. Soc., Los Angeles, pp. 57–64.

Karatsuba, A. A., and Ofman, Yu. P., 1962, Multiplication of multidigit numbers on automata, *Dokl. Acad. Nauk SSSR,* **145** (2), 293–294. (Translated in *Soviet Phys. Dokl.,* **7**, 595–596.)

Karp, R. M., 1972, Reducibility among combinatorial problems, in *Complexity of Computer Computations,* R. Miller and J. Thatcher, eds., Plenum Press, New York, pp. 82–104.

Khachian, L. C., 1979, A polynomial time algorithm for linear programming, *Dokl. Akad. Nauk SSSR,* **2444** (5), 1093–1096. (Translated in *Soviet Math. Doklady,* **20**, 191–194.)

Klee, V., and Minty, G. J., 1972, How good is the simplex algorithm?, in *Inequalities,* vol. III, O. Shisha, ed., Academic Press, New York, pp. 159–175.

Kolmogorov, A. N., 1953, On the notion of an algorithm, *Uspekhi Math. Nauk*, **8**, 175–176.

Kolmogorov, A. N., and Uspenskii, V. A. 1963, On the definition of an algorithm, *Am. Math. Soc. Transl.*, **II** (29), 217–245.

Knuth, D. E., 1968, *The Art of Computer Programming*, vol. 1, Addison-Wesley, Reading, Mass.

Knuth, D. E., 1973, *The Art of Computer Programming*, vol. 3, Addison-Wesley, Reading, Mass.

Knuth, D. E., and Szwarcfiter, J. L., 1974, A structured program to generate all topological sorting arrangements, Inform. *Processing Letters*, **2**, 153–157.

Kreczmar, A., 1976, On memory requirements of Strassen algorithms, in *Algorithms and Complexity: New Directions and Recent Results*, J. F. Traub, ed., Academic Press, New York.

Kruskal, C. P., 1983, Searching, merging, and sorting in parallel computation, IEEE Trans. Comp., c-32, 10, pp. 942–946.

Kung, H. T., 1976, Synchronized and asynchronous parallel algorithms for multi-processes, in *Algorithms and Complexity: New Directions and Recent Results*, J. F. Traub, ed., Academic Press, New York.

Laderman, J. D., 1976, A non-commutative algorithm for multiplying 3×3 matrices using 23 multiplications, *Bull. Am. Math. Soc.*, **82**, 126–128.

Lee, S., 1984, M.Sc. project dissertation, The Centre for Computing and Computer Science, Univ. of Birmingham, Birmingham, England.

Lenstra, A. K., Lenstra, H. W., and Lovatz, L., 1982, Factoring polynomials with rational coefficients, *Rep. 82–05*, Univ. of Amsterdam, Dept. of Math.

Levin, L. A., 1973, Universal search problems, *Problemy Peredaci Informacii*, **9**, 115–116. (Translated in *Problems of Information Transmission*, **9**, 265–266.)

Levin, M., 1984, M. Sc. project dissertation, the Centre for Computing and Computer Science, Univ. of Birmingham, Birmingham, England.

Levy, H., and Lessman, F., 1961, *Finite Difference Equations*, Macmillan, New York.

Lin, S., 1965, Computer solution of the travelling salesman problem, *Bell System Tech. J.*, **44**, 2245–2269.

Lint, B., and Agerwala, T., 1981, Communication issues in the design and analysis of parallel algorithms, *IEEE Trans. Softw. Eng.*, **SE–7**, 174–188.

Little, J. D. C., Murt, K. G., Sweeney, D. W., and Karel, C., 1963, An algorithm for the travelling salesman problem, *Operations Research*, **11**, 972–989.

Liu, C. L. 1968, *Introduction to Combinatorial Mathematics*, McGraw-Hill, New York.

Lueker, G. S., 1980, Some techniques for solving recurrences, *Computing Surveys*, **12** (4), 418–436.

Luks, E. M., 1980, Isomorphism of graphs of bounded valence can be tested in polynomial time, *Proc. 21st IEEE Symp. on Foundations of Computer Science*, IEEE Comp. Soc., Los Angeles, pp. 42–49.

Matijasevich, Y. V., 1970, Enumerable sets are diophantine, *Dokl. Akad. Nauk SSSR*, **191**, 279–282 (in Russian). English translation in *Soviet Math. Dokl.*, **11**, 354–357.

Meyer, A. R., and Stockmeyer, L. J., 1972, The equivalence problem for regular expressions with squaring requires exponential time, *Proc. 13th Ann. Symp. on Switching and Automata Theory*, IEEE Computer Soc., Long Beach, Calif, pp. 125–129.

Miller, G. L., 1975, Reimann's hypothesis and test for primality, *Proc. of the Seventh Annual ACM Symp. on Theory of Computing*, pp. 234–239.

Mukai, H., 1979, Parallel algorithms for solving systems of nonlinear equations, *Proc. 17th Annual Allerton Conf. on Comm., Control and Comput.*, 10–12 Oct., pp. 37–46.

217

Muller, D. E., and Preparata, F. P., 1975, Bounds to complexities of networks for sorting and for switching, *J ACM*, **22**, 195–201.

Munro, J. I., 1977, The parallel complexity of arithmetic computation fundamentals of computation theory, *Lecture Notes in Comp. Sc. 56*, Springer-Verlag, pp. 466–475.

Nassimi, D., and Sahni, S., 1979, Parallel permutation and sorting algorithms, *Proc. 7th Annual Allerton Conf. on Comm., Control and Comput.*, 10–12 Oct., pp. 1–10.

Norrie, C., 1984, Supercomputers for superproblems: an architectural introduction, *Computer*, IEEE Computer Soc., **17** (3), 62–74.

Orenstein, J. A., Merrett, T. H., and Devroye, L., 1983, Linear sorting with $O(\log n)$ processors, *BIT*, **23** (2), 170–180.

Ortega, J. M. and Rheinboldt, W. C., 1970, *Iterative Solution of Nonlinear Equations in Several Variables*, Academic Press, New York.

Pacault, J. F. 1974, Computing the weak components of a direct graph, *SIAM J. Comput.*, **3**, 56–61.

Pan, V., 1972, On schemes for the computation of products and inverses of matrices, *Russian Math. Surveys*, **27** (5), 249–250.

Pan, V., 1978, Strassen algorithm is not optimal. Trilinear technique of aggregating, uniting and cancelling for constructing fast algorithms for matrix multiplication, *Proc. 19th Annual Symposium on the Foundations of Computer Science*, Ann Arbor, MI, pp. 166–176.

Pan, V., 1980, New fast algorithms for matrix operations, *SIAM J. Comput.*, **9** (2), 321–342.

Parkynson, D., 1982, Using the ICL DAP, *Comput. Physics Comm.*, **26**, 227–232.

Pease, M. C., 1977, The indirect binary n-cube microprocessor array, *IEEE Trans. Comput.*, **C–26**, 458–473.

Pippenger, N., 1979, On simultaneous resource bounds (preliminary version). *Proc. 20th IEEE Symp. on Foundations of Computer Science*, IEEE Computer Society, Los Angeles, pp. 307–311.

Pippenger, N., and Fischer, M. J., 1979, Relations among complexity measures, *JACM*, **26**, 2, pp. 361–381.

Polak, E., 1974, A globally converging secant method with applications to boundary value problems, *SIAM J. on Num. Analysis*, **11**, 529–537.

Preparata, F. P., 1978, New parallel sorting schemes, *IEEE Trans. Comput.*, **C–27**, 667–673.

Preparata, F., and Vuillemin, J., 1979, The cube-connected cycles: a versatile network for parallel computation, *20th Annual Symp. on Foundations of Comput. Sci.*, Puerto Rico, pp. 140–147.

Pressburger, M., 1929, Über die vollständigkeit eines gewissen systems der arithmetik ganzer zahlen, in welchen die addition als einzige operation hervortritt, *Comptes Rendus. I Congrès des Math. des Pays Slaves*, Warsaw, pp. 92–101.

Purcell, C. J., 1982, Using the Cyber 205, *Comput. Physics Comm.*, **26**, 249–251.

Rabin, M. O., 1960, Degree of difficulty of computing a function and a partial ordering of recursive sets, *Tech. Rep.*, *1*, ONR, Jerusalem.

Rabin, M. O., 1976, Probabilistic algorithms, in *Algorithms and Complexity: New Directions and Recent Results*, J. F. Traub, ed., Academic Press, New York, pp. 21–39.

Radon, J., 1917, Über die bestimmung von functionen durch ihre integralwerte langs gewisser mannigfaltigkeiten, *Ber. Verh. Sachs. Akad. Wiss. Leipzig, Math–Nature*, Kl. 69.

Ramakrishnan, I. V., and Browne, J. C., 1983, A paradigm for the design of parallel algorithms with applications, *IEEE Trans. Softw. Eng.*, **SE–9** (4), 411–415.

Reingold, E., Nevergelt, J., and Deo, N., 1977, *Combinatorial Algorithms: Theory and Practice*. Prentice-Hall, New York.

Rivest, R. L., Shamir, A., and Adleman, L., 1978, A method for obtaining digital signatures and public-key cryptosystems, *Comm. ACM,* **21** (2), 120–126.

Rivest, R., and Vuillemin, J., 1975, A generalization and proof of the Andreaa-Rosenburg conjecture, *Proc. Seventh Annual ACM Symp. on Theory of Computing,* ACM, New York, pp. 6–11.

Rogers, H. Jn., 1967, *Theory of Recursive Functions and Effective Computability,* McGraw-Hill, New York.

Rose, D. J., Tarjan, R. E., and Lueker, G. S., 1976, Algorithmic aspects of vertex elimination on graphs, *SIAM J. Comput.,* **5**, 266–283.

Rosenfeld, J., and Driscoll, G. C., 1969, Solution of the Dirichlet problem on a simulated parallel processing system, in *Information Processing,* vol. 68, North-Holland, Amsterdam, pp. 499–507.

Rosenstiehl, P. *et al.,* 1972, Intelligent graphs: networks of finite automata capable of solving graph problems, in *Graph Theory and Computing,* R. Read, ed., Academic Press, New York.

Russell, R. M., 1982, The Cray–1 computer system, in *Computer Structures: Principles and Examples,* Siewiorek, D., Bell, G., and Newell, A., Eds., McGraw-Hill, Hightstown, New York, pp. 743–752.

Schachtel, G., 1978, A non-commutative algorithm for multiplying 5×5 matrices using 103 multiplications, *Information Processing Lett.,* No. 4, 180–182.

Schönauer, W., 1983, The efficient solution of large linear systems, resulting from the FDM for 3-D PDE7s, on vector computers, *EDF.-Bull. de la Direction des Etudes et des Recherches,* Série C, **1**, 135–142.

Schönhage, A., 1973, Real-time simulation of multidimensional Turing machines by storage modification machines, *Project MAC Tech. Memorandum 37,* Massachusetts Institute of Technology, Cambridge, Mass.

Schönhage, A., 1980, Storage modification machines, *SIAM J. Comput.,* **9**, 490–508.

Schönhage, A., Paterson, M., and Pippenger, N., 1975, Finding the median, *J. Comp. Syst. Sci.,* **13**, 184–199.

Schönhage, A., and Strassen, V., 1971, Schelle multiplication grosser zahlen, *Computing,* **7**, 281–292.

Sethi, R., 1975, Complete register allocation problems, *SIAM J. Comput.,* **4**, 226–248.

Shamanskii, V. E., 1967, On a modification of Newton's method, *Ukrain. Mat. Zh.,* **19**, 133–138.

Solovay, R., and Strassen, V., 1977, A first Monte Carlo test for primality, *SIAM J. Comput.,* **6**, 84–85.

Sorenson, P. G., Tremblay, J. P., and Deutscher, R. F., 1978, *Key-to-address transformation techniques,* INFOR 16, pp. 1–34.

Stanley, R. P., 1978, Generating functions, in *MAA Studies in Mathematics, 17: Studies in Combinatorics,* Gian-Carlo Rota, ed., the Mathematical Association of America, pp. 100–141.

Stearns, R. E., Hartmanis, J., and Lewis, P. M., 1965, Hierarchies of memory limited computations, *6th IEEE Symp. on Switching Circuit Theory and Logical Design,* pp. 179–190.

Stockmeyer, L. J., 1977, The polynomial-time hierarchy, *Theoretical Comput. Sci.,* **3**, 1–22.

Stockmeyer, L. J., 1979, Classifying the computational complexity of problems, *IBM Research Report, RC 7606* (no. 32926), IBM Thomas J. Watson Research Center, Yorktown Heights, NY 10598.

Stone, H. S., 1971, Parallel processing with perfect shuffle, *IEEE Trans. Comput.,* **C–20**, 153–161.

Strassen, V., 1969, Gaussian elimination is not optimal, *Numer. Math.,* **13**, 354–356.

Strassen, V., 1972, Evaluation of rational functions, in *Complexity of Computer Computations,* R. E. Miller and J. W. Thatcher, eds., Plenum Press, New York, pp. 1–10.

Tarjan, R. E., 1972, Depth-first search and linear graph algorithms, *SIAM J. Comput.*, **1**, 146–160.

Tarjan, R. E., 1974, A new algorithm for finding weak components, *Information Processing Lett.*, **3**, 13–15.

Tarjan, R. E., 1975, Solving path problems on directed graphs, *Tech. Rep. STAN-CS-75-528*, Comp. Sci. Dept., Stanford Univ., Stanford, Calif.

Tarjan, R. E., 1977, Reference machines require non-linear time to maintain disjoint sets, *Proc. Ninth Annual ACM Symp. on Theory of Computing*, ACM, New York, pp. 18–29.

Tarjan, R. E., 1978, Complexity of combinatorial algorithms, *SIAM Review*, **20**, 457–491.

Tarjan, R. E., 1983, Data Structures and Network Algorithms, *CBMS-NSF Regional Conf. Series in Appl. Math.*, SIAM, Philadelphia, Pennsylvania 19103.

Taylor, D. J., and Hopkinson, J. F. L., 1982, The Cray–1S and the Cray service provided by the SERC at the Daresbury Lab., *Comput. Physics Comm.*, **26**, 259–265.

Temperton, C., 1983, Fast Fourier transform for numerical prediction models on vector computers, *EDF.-Bull. de la Direction des Etudes et des Recherches*, Série C, **1**, 159–162.

Thompson, C. D., and Kung, H. T., 1977, Sorting on a mesh-connected parallel computer, *Comm. ACM*, **20** (4), 263–271.

Todd, S., 1978, Algorithm and hardware for a mergesort using multiple processors, *IBM J. Research Development*, **22**, 509–517.

Traub, J. F., 1964, *Iterative Methods for the Solution of Equations*, Prentice-Hall, Englewood Cliffs, N.J.

Turing, A. M., 1936, On computable numbers with an application to the entscheidnungsproblem, *Proc. London Math. Soc.*, Ser. 2, **42**, 230–265; A correction, *ibid.*, **43**, 544–546.

Valiant, L. G., 1975, Parallelism in comparison problems, *SIAM J. Comput.*, **4**, 348–355.

Valiant, L. G., 1979a, The complexity of enumeration and reliability problems, *SIAM J. Comput.*, **8**, 410–421.

Valiant, L. G., 1979b, The complexity of computing the permanent, *Theoretical Comp. Sci.*, **8**, 189–202.

Varol, Y. L., and Rotem, D., 1981, An algorithm to generate all topological sorting arrangements, *The Computer Journal*, **24**, 83–84.

Vuillemin, J., 1978, A data structure for manipulating priority queues, *Comm. ACM*, 21, pp. 309–314.

Wah, B. W., 1984, File placement on distributed computer systems, Computer, *IEEE J. Comput.*, **17** (1), 23–32.

Winograd, S., 1971, On multiplication of 2 × 2 matrices, *Linear Algebra and Applications*, **4**, 381–388.

Winograd, S., 1980, *Arithmetic Complexity of Computations*, Society for Industrial and Applied Mathematics, Philadelphia, PA 19103.

Wirth, N., 1984, Data structures and algorithms, *Scient. American*, **251**, 3, 48–57.

Wolfe, P., 1959, The secant method for simultaneous nonlinear equations, *Comm. ACM*, **2**, 12–14.

Young, P., 1973, Easy construction in complexity theory: gap and speed-up theorems, *Proc. Am. Math. Soc.*, **37**, 555–563.

Table of Notations

$\lceil x \rceil$	the least integer greater than or equal to x
$\lfloor x \rfloor$	the greatest integer less than or equal to x
$\mathbb{N}$	the set of all natural numbers
$x(\bmod y)$	mod function, i.e. if $y = 0$ then 0 else $x - y\lfloor x/y \rfloor$
$a \equiv x(\bmod y)$	relation of congruence, i.e. $a = x + yt$, $t \in \mathbb{N}$
$x := y$	y defines x
$\begin{pmatrix} x \\ n \end{pmatrix}$	binomial coefficient
J or $J(x)$	the Jacobian matrix
$f(n) = 0(g(n))$	large 0 notation, i.e. $f(n)/g(n)$ is bounded as $n \to \infty$
$A \cup B$	union of the sets A and B
$A \cap B$	intersection of the sets A and B
$P \subseteq Q$	P is a subset of Q
$\forall$	for all
$\exists$	there exists at least
$\wedge$	conjunction
$\vee$	disjunction
Δ	the null value of a pointer

Index

222